群胚和弱乘子 Hopf 代数

王栓宏 著

科学出版社

北京

内 容 简 介

本书主要介绍了群胚、拟群、箭图、乘子、环扩张以及与之相关的各类 Hopf 型代数的基本概念和理论，尤其讨论了弱乘子 Hopf 代数概念的发展、构造以及对偶理论. 本书内容由浅入深，既有理论又有新的应用，反映了近十几年来在代数量子群(胚)理论中国际最新的研究成果，是国内外反映该研究领域的专著之一.

本书可供大学数学和数学物理专业的高年级本科生、硕士与博士研究生、高校教师以及研究工作者阅读和参考.

图书在版编目(CIP)数据

群胚和弱乘子 Hopf 代数/王栓宏著. —北京：科学出版社，2018.2
ISBN 978-7-03-056659-1

I. ①群··· Ⅱ. ①王··· Ⅲ. ①广群 ②Hopf 代数 Ⅳ. ①O152

中国版本图书馆 CIP 数据核字 (2018) 第 039879 号

责任编辑：李 欣／责任校对：邹慧卿
责任印制：吴兆东／封面设计：陈 敬

科学出版社 出版
北京东黄城根北街 16 号
邮政编码：100717
http://www.sciencep.com

北京厚诚则铭印刷科技有限公司 印刷
科学出版社发行 各地新华书店经销
*

2018 年 2 月第 一 版　　开本：720 × 1000 B5
2024 年 2 月第四次印刷　印张：11 1/2
字数：232 000
定价：79.00 元
(如有印装质量问题，我社负责调换)

前　　言

一个有限群上的复函数代数构成一个 Hopf 代数[2, 173], 但当群是无限时, 不再是 Hopf 代数而是乘子 Hopf 代数[189, 190]; 当考虑群胚时, 一个有限群胚上的复函数代数构成一个弱 Hopf 代数[25], 但当群胚是无限时, 它不再是弱 Hopf 代数, 而是弱乘子 Hopf 代数[199-202].

另一方面, 有点 Hopf 代数的分类问题最辉煌的成果是 N. Andruskiewitsch 和 H. J. Schneider 在 2010 年发表在 *Ann. Math.* 上的文章[13], 主要用到 Nichols 的代数提升理论. Nichols 代数是一个辫子张量范畴中的 Hopf 代数 (即辫子 Hopf 代数), 它包含着某些泛性质且自然出现在每个有点 Hopf 代数中. 最有名的例子是小量子群的 Borel 部分 $u_q(\mathfrak{g})^{\pm}$. 事实上, 一般它也保留着 Lie-理论的成分, 且任何 Nichols 代数由一个 Weyl 群胚所控制[10]. 由此可见, 群胚的研究无论是对弱乘子 Hopf 代数还是对有点 Hopf 代数分类都具有很重要的现实意义.

人们[76,129,171,177,179-181,262,265] 对著名的 Pontryagin 对偶定理的各种推广的构造理解产生了算子代数的方法来逼近量子群; 其历史比现在人们所理解的量子群[72, 92, 127] 早.

我们可以回顾一下, 已知一个局部紧交换群 G, 它的对偶群 $\hat{G}$ (即由 G 上的连续特征函数构成的群) 也是局部紧交换群, 那么 Pontryagin 对偶定理是说对偶群 $\hat{G}$ 的再次对偶结果自然同构于原来的群 G[172]. 这一定理是抽象的调和分析 (即扩充了的傅里叶级数与傅里叶积分理论) 的基础.

从代数和分析的角度, 量子群首次出现在 G.I.Kac[94] 关于群环的研究工作中, 解决的问题是 Pontryagin 对偶扩张到非交换群的情况. 因为一般的非交换群的对偶不再是非交换群, 所以我们需要一个包含局部紧群及其对偶的类. 例如, 无限维 Hopf 代数或量子群的对偶不再是 Hopf 代数或量子群, 那么尤其是无限维群代数的对偶是什么?

为了解决以上问题, 1994 年, 比利时著名的数学物理学家、算子代数学家 A. Van Daele 引进了乘子 Hopf 代数[189] 的概念, 该概念利用算子的方法抽象简单地推广了一般的 Hopf 代数. 同时, 在 1996 年, 建立在积分概念基础之上, 他用代数的方法定义了一类特殊的量子群, 称之为代数量子群[190], 它包含了所有的紧型和离散型量子群. 粗略地说, 一个代数量子群就是一个带有正的不变函数 (Haar 测度) 的乘子 Hopf*-代数, 这样一类代数体系的两次对偶仍然在该类中, 而且与原来的代数体系同构, 也就是说, Pontryagin 对偶定理成立. 随后, 为了寻求

Pontryagin 对偶定理成立的其他范畴, 相继出现了广义余-Frobenius 代数[41]、代数量子超群[56, 57] 和有界型量子 (超) 群理论[198, 209, 233, 234, 288, 289].

作者 2004 年在比利时鲁汶天主教大学做博士后研究工作时, 向 A. Van Daele 教授研习乘子 Hopf 代数理论, 并分别于 2006 年和 2009 年进一步进行合作交流, 在代数量子群方面取得了一些国际领先的研究成果. 此外, 作为作者博士后期间的一个课题, 从 2004 年到 2009 年, 作者对无限群胚代数及其对偶作了研究, 但这个问题远远超过我们预期的研究进展, 研究工作非常困难, 因为涉及弱 Hopf 代数、乘子代数和李理论. 但幸运的是, 在历经 6 年研究之后, 作者终于得到了一个比较满意的定义, 称之为 "弱乘子 Hopf 代数"[199–202], 作为本书的主要贡献. 同时, 第 5 章给出了许多有待研究的新课题. 我们主要阐述了研究这一理论的动机、思想方法和如何分析一些假设条件的来源. 在为了得到弱乘子 Hopf 代数的一个好的定义过程中, 读者可以体会到作者的研究思路和如何发展一个新理论的方法. 但这方面只是一个开始, 有待激发读者的研究热情.

本书的目的是继续介绍国际前沿学科的研究方向: (弱) 代数量子 (超) 群以及 Hopf 型代数. 读者可以从书中领略到这一理论具有很强的概括性、处理问题简明、涉及面广的特点. 本书的取材具有很深的数学分析、算子代数及其物理背景, 是建立在作者近年来与同行专家合作研究的基础之上的产物. 在写作方面, 本书尽量做到自成体系.

书中难免有不足之处, 恳请读者批评指正.

在完成本书时, 作者首先要感谢导师许永华教授在硕士与博士研究生阶段对作者的谆谆教导. 感谢 A. Van Daele 教授在作者做博士后期间以及多年来的合作访问中所给予的各种帮助. 还要感谢阿根廷科尔多瓦大学数学与物理系 N. Andruskiewitsch 教授研究团队、美国印地安那大学数学系 V. Turaev 教授以及韩国全北国立大学数学系 Y. G. Kim 教授, 在作者合作访问期间给予的大力帮助. 作者也借此机会感谢他的妻子与女儿这么多年来无怨无悔的大力支持.

本书得到了东南大学数学系和国家自然科学基金项目 (No. 11371088) 的资助.

在完成本书之际, 作者以一首诗表达这么多年来进行研究的一点体会:

善于思考找问题, 苦思冥想不放弃,
专研过程苦为乐, 解决问题不知疲;
劳逸结合思道理, 归纳分析作对比,
坚持不懈恒心在, 体系建立终有日.

作 者

2016 年 10 月于南京

目　　录

前言
第 1 章　基本概念 ... 1
　1.1　群胚 ... 1
　1.2　拟群 ... 10
　1.3　箭图 ... 12
　1.4　乘子 ... 13
　1.5　环扩张 ... 15
　1.6　辫子与纽结 ... 19
第 2 章　Hopf 代数 ... 25
　2.1　双代数的范畴描述 ... 25
　2.2　Hopf 代数定义 ... 29
　2.3　积分 ... 39
　2.4　对偶 ... 42
　2.5　一类诱导的群胚 ... 46
第 3 章　弱 Hopf 代数 ... 53
　3.1　弱 Hopf 代数定义 ... 53
　3.2　积分性质 ... 60
　3.3　Drinfel'd (余) 偶 ... 63
　3.4　深度 2 扩张 ... 67
　3.5　对偶 ... 70
第 4 章　Hopf 型代数 ... 75
　4.1　(弱) 左 (右)Hopf 代数 ... 75
　4.2　(弱) 双 Frobenius 代数 ... 76
　4.3　(弱)(余) 拟 Hopf 代数 ... 78
　4.4　(弱) 拟 Hopf 群 (余) 代数 ... 81
　4.5　(弱)Hom-Hopf 代数 ... 84
　4.6　(弱)Hopf (余) 拟群 ... 87
　4.7　Hopf 代数胚 ... 89
　4.8　其他 Hopf 代数系 ... 96

第 5 章　弱乘子 Hopf 代数 ... 101
5.1　定义与例子 ... 101
5.2　弱乘子双代数 ... 108
5.3　Pontryagin 对偶 ... 111
5.4　乘子 Hopf 代数胚 ... 122
5.5　进一步研究的问题 ... 129

第 6 章　非线性方程与微积分 ... 135
6.1　非线性方程 ... 135
6.2　Hopf 代数方法 ... 146
6.3　进一步研究问题 ... 152

第 7 章　范畴中的基本概念 ... 153
7.1　范畴与函子 ... 153
7.2　辫子张量范畴 ... 157
7.3　张量积 ... 160

参考文献 ... 162
名词索引 ... 177

第1章 基本概念

本章主要介绍群胚 (groupoid)、拟群 (quasigroup)、箭图 (quiver)、辫子 (braid) 和纽结 (knot) 与环扩张等概念, 它们的深层次研究会涉及 (弱) Hopf 代数 ((weak) Hopf algebra)[13,25,108,109] 与弱乘子 Hopf 代数 (weak multiplier Hopf algebra) 理论的建立以及 Hopf 代数 (胚) 表示理论 [73, 168, 187, 203, 205, 280].

1.1 群 胚

本节提供群胚 [11, 27, 199, 290] 的两种常用的定义: 第一个是看做群定义的推广的代数定义法, 第二个是小范畴定义法.

定义 1.1.1 一个**群胚**是一个非空集合 G 带有一个一元运算 $^{-1}: G \longrightarrow G$ 和一个部分二元运算 $*: G \times G \longrightarrow G$, 使得下面条件满足: 对任意 $f, g, h \in G$,

(i) **存在性** $(f*g)*h$ 有定义当且仅当 $f*g$ 和 $g*h$ 有定义.

(ii) **结合性** $(f*g)*h$ 有定义当且仅当 $f*(g*h)$ 有定义且 $(f*g)*h = f*(g*h)$.

(iii) **单位性** 存在唯一元素 $s(g), t(g) \in G$ 使得 $gs(g)$ 和 $t(g)g$ 有定义且 $gs(g) = g = t(g)g$.

(iv) **可逆性** 存在唯一元素 $g^{-1} \in G$ 使得 $s(g) = g^{-1}*g$ 和 $t(g) = g*g^{-1}$.

设 G 是一个群胚. 当 G 是有限集合时, 我们说 G 是**有限群胚**; 当 G 是无限集合时, 我们说 G 是**无限群胚**.

注记 1.1.2 对任意 $g, h \in G$, 有:

(i) 从定义不难推出: $(g^{-1})^{-1} = g$ 且 $g*h*h^{-1} = g$ 和 $g^{-1}*g*h = h$.

(ii) $g*h$ 有定义当且仅当 $s(g) = t(h)$ 当且仅当 $h^{-1}*g^{-1}$ 有定义. 此时, 有 $(g*h)^{-1} = h^{-1}*g^{-1}$, $s(gh) = s(h)$, $t(gh) = t(g)$. 记 $G^{(2)} = \{(g,h) \in G \times G \mid s(g) = t(h)\}$.

(iii) 群胚 G 可以简单地说, 它是一个非空集合 G 带有一个乘积, 该乘积不是对于所有一对元素 $g, h \in G$ 都有定义 gh. 只有当元素 g 的所谓的**源**(source): $s(g)$ 等于元素 h 的所谓的**靶**(target): $t(h)$ 时, 才有乘积 $gh \in G$. 源与靶是从 G 到 G 的所谓的单位 (即逆元) 集合的映射. 这里单位集合是 G 的一个子集合. 乘积满足结合性, 且对任意 $g \in G$, 存在唯一逆元 g^{-1}, 这里逆元被刻画为: $s(g^{-1}) = t(g)$ 和 $t(g^{-1}) = s(g)$ 且满足: $s(g) = g^{-1}g$ 和 $t(g) = gg^{-1}$.

(iv) 在文献中, **源**也称为 "**定义域**"(domain) 或 "**头**" (head) 或 "**源头**" (origin); 类似地, **靶**也称为 "**值域**" (range) 或 "**尾**" (tail) 或 "**末端**" (end).

(v) 群胚 G 的**反群胚** (opposite groupoid) 记为 G^{op}, 它的源是 G 的靶且它的靶是 G 的源 (见定义 1.1.3).

(vi) 一个元素 $e \in G$ 称为 G 的一个**单位元** (identity), 如果 $e = s(g) = t(g^{-1})$, 对 $g \in G$. 在此情况下, e 称为是 g 的源单位元和 g^{-1} 的靶单位元. 记 G_0 是 G 中所有单位元集合. 对于 $e \in G_0$, 有 $s(e) = t(e) = e, e^{-1} = e$, 且记 $G_e = \{g \in G \mid s(g) = t(g) = e\}$. 显然, G_e 是一个群, 称为是结合 e 的**迷向** (isotrop) 群或**主** (principal) 群.

(vii) 设 G 是一个群胚. 对任意 $u, v \in G_0$, 设 $G^u = t^{-1}\{u\}, G_v = s^{-1}\{v\}$ 和 $G_v^u = G^u \cap G_v$.

(viii) 我们说 G 是一个**拓扑群胚**(topological groupoid), 如果 G 上的逆映射 $x \mapsto x^{-1}$ 和从 $G^{(2)}$ 到 G 的乘积映射 $(x, y) \mapsto xy$ 都是连续映射.

下面我们给出群胚的第二个等价定义.

定义 1.1.3 群胚 G 通常被看成是一个**小范畴**(small category) 使得每个态射 (morphism) 或箭 (arrow) 可逆或同构. 这个等价于下面数据: 设 G 和 B 是两个非空集合. 一个**基**(base)B(或对象集合) 上的一个**群胚**G(或态射集合) 是一个带有**源投射** s, 一个**靶投射** t: $s, t: G \longrightarrow B$, 一个**内射** (也称恒等态射)$E: B \longrightarrow G$, 一个**乘法** (也称态射合成) $m: \{(g, h) \in G \times G \mid t(g) = s(h)\} \longrightarrow G$, 记乘积 $m(g, h) = h \circ g \in G$ 和一个**卷积** $i: G \longrightarrow G$ 使得 $s(i(x)) = t(x), t(i(x)) = s(x), m(i(x), x) = \mathrm{Id}_{s(x)}$ 和 $m(x, i(x)) = \mathrm{Id}_{t(x)}$ 对任意 $x \in G$, 满足通常的运算. 我们可以把 G 中元素看成是箭 $s(g) \xrightarrow{g} t(g)$. 当 $|B| = 1$ 时, G 是一个群. 设 $p, q \in B$, 像通常一样, 记 $G(p, q)$ 是从 p 到 q 的所有箭的集合, 且 $G(q) = G(q, q)$.

我们说 G 是**李群胚**如果 B, G 都是光滑流行 (smooth manifolds), 所有结构映射是光滑的且 s, t 都是子侵入 (submersion).

注记 1.1.4 (i) 设 B 是一个非空集合. 已知两个映射 $p: X \longrightarrow B$ 和 $q: Y \longrightarrow B$. 设

$$X_p \times_q Y = \{(x, y) \in X \times Y \mid p(x) = q(y)\}.$$

那么由定义 1.1.3, 一个反群胚 G^{op} 是由一个箭的集合 G, 对象 (object) 集合 B(称为 "基"), 源和靶映射 $s, t: G \longrightarrow B$, 合成 $m: G_s \times_t G \longrightarrow G$ 与单位元 $\mathrm{id}: B \longrightarrow G$ 组成. 通常用 G 表示群胚和箭的集合. 记 $m(f, g) = fg$ (不是 gf), 对 $(f, g) \in G_s \times_t G$, 且把 B 与 G 的子集合 $\mathrm{id}(B)$ 看成一样.

(ii) 群胚 G 也可以简单描述为: 它是一个非空集合 G 带有一个部分结合乘法和部分单位 (units), 它的元素都是可逆的. 源和靶映射被确定为 $s(g) = g^{-1}g$ 和 $t(g) = gg^{-1}$.

1.1 群胚

(iii) 群胚之间的态射是一个保持乘积的映射,因此也保持源与靶并且诱导基之间的映射. 如果两个群胚 G 和 H 有相同的基 P,我们说 $T: G \longrightarrow H$ 是 P 上的群胚同态如果限制映射 $P \longrightarrow P$ 是恒等映射. P 上的群胚构成一个范畴 $\text{Gpd}(P)$.

(iv) 一个宽子群胚 (wide subgroupoid) 是一个子群胚使得包含是 P 上的群胚态射. 一个群胚的子群丛 (bundle) 是一个由群丛构成的子群胚. 存在群胚 G 的一个最大的宽的子群丛, 即 $G^{\text{bundle}} = \coprod_{p \in P} G(p)$, 亦即我们忘记不同点之间的箭.

(v) 如果 $(N_i)_{i \in I}$ 是群胚 G 的一簇宽的子群胚, 那么 $\bigcap_{i \in I} N_i$ 是 G 的一个宽的子群胚, 其定义为 $\bigcap_{i \in I} N_i(p,q) = \bigcap_{i \in I} (N_i(p,q))$.

(vi) 有时, 对象的集合记为 G^0, 态射的集合记为 G^1. 作为代数定义, 态射集合 G^1 被看成是 G 的元素, 所以, 从某种程度上说, G 可以等同于集合 $G(x,y)$ 的非交并, 对任意元素 $x, y \in G^0$.

(vii) 最初的文献中, 一般对象称为**地方**(place), 态射称为**路**(road). 如果 $\alpha: x \longrightarrow y, \beta: y \longrightarrow z$ 属于 G^1, 则它们的合成写为 $\alpha + \beta: x \longrightarrow z$, 同时, α 的逆记为 $-\alpha: y \longrightarrow x$. 群胚 G 的**子群胚**(subgroupoid) H 是 G 的一个子范畴, 使得 H 也是一个群胚. 这个条件简单来说, 如果 α 是 H 的路, 那么 $-\alpha$ 也是 H 的路. 一个子群胚 H 是**满的**(full), 如果对任意两个地方 $x, y \in H$, 从 x 到 y 的路集合 $H(x,y)$ 和 $G(x,y)$ 相等. G 的任意地方集合决定一个满的子群胚. 尤其, 对一个单个地方 $x \in G$, 这个满的子群胚是一个群, 称为在 x 处的**地方群**(place group) 或**顶点群**(vertex group). 一个群胚 G 是**连通的**(connected) 如果对任意地方 x, y, 在 G 中存在从 x 到 y 的一条路. 一个群胚 G 的**分支**(component) 是最大的连通的 G 的子群胚. 容易看出, 这些分支存在而且是满的子群胚. 如果 G 的分支都是地方群, 那么 G 称为**全非连通的**(totally disconnected). 群胚 G 的子群胚 H 称为是**表示**(representative) 如果对每个地方 $x \in G$ 存在从 x 到 H 的一个地方的一条路. 因此, H 是表示如果 H 与 G 的每个分支相交. 设 G, H 是群胚. 一个态射 $f: G \longrightarrow H$ 是一个共变 (covariant) 函子. 因此, 对任意地方 $x \in G$, 有地方 $f(x) \in H$, 且在 G 中对任意 $\alpha: x \longrightarrow y$, 在 H 中有 $f(\alpha): f(x) \longrightarrow f(y)$, 而且, 当 $\alpha + \beta$ 有定义, f 满足: $f(\alpha + \beta) = f(\alpha) + f(\beta)$. 设 $f, g: G \longrightarrow H$ 群胚态射. 一个**同伦**(homotopy)$\theta: f \simeq g$ 简单地为 f, g 的一个自然等价, 当 f, g 被看成函子时. 因此, 对地方 $x \in G$, 有一条路 $\theta(x): f(x) \longrightarrow g(x)$, 使得对任意 $\alpha: x \longrightarrow y$, 有下面交换图:

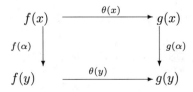

即 $f(\alpha) + \theta(y) = \theta(x) + g(\alpha)$. 两个群胚 G, H 称为等价如果存在同态 $f : G \longrightarrow H$ 和 $g : H \longrightarrow G$ 使得 $gf \simeq 1, fg \simeq 1$.

事实上, 任何群胚等价于在每个分支中的一个地方群组成的全非连通群胚. 设 H 是 G 的子群胚, $i : H \longrightarrow G$ 是包含映射. 从 G 到 H 的一个满射**收缩**(retraction) 是一个态射 $r : G \longrightarrow H$ 使得 $ri = 1$, 如此的一个收缩是一个扭曲收缩 (deformation retraction) 如果 $ri = 1$ 和 $ir \simeq 1$. 如果这样的 r 存在, 我们分别称 H 是 G 的收缩 (retract) 和扭曲收缩.

G 的子群胚 N 称为**正规的**(normal), 如果 N 和 G 有相同的地方, 且: ① $N(x, y) = \varnothing, x \neq y$, ② $\alpha + N(y, y) - \alpha \subseteq N(x, x)$ 对任意 $\alpha \in G(x, y)$ 和所有地方 $x, y \in G$. 如此的正规子群胚决定一个商群胚 G/N.

例 1.1.5 (i) 设 G_1 和 G_2 是分别带有单位元 e_1 和 e_2 的两个群. 那么 $G = G_1 \cup G_2$ 是一个群胚. 进一步, 有 $G^{(0)} = \{e_1, e_2\}$, $s(x) = t(x) = e_1$ 和 $s(y) = t(y) = e_2$ 对任意 $x \in G_1$ 和 $y \in G_2$. 更一般地, 设 G 是无限多个群 $G_1 = G_2, \cdots, G_i, \cdots$ 的非交并, 那么 G 是一个群胚.

(ii) 设 $\sim$ 是一个非空集合 X 上的一个等价关系. 设 $G = \{(y, x) \mid x \sim y, x, y \in X\}$. 那么 $G^{(0)} = \{(x, x) \mid x \in X\} \equiv \{x \mid x \in X\} = X$, $s(y, x) = (x, x) \equiv x$ 和 $t(y, x) = (y, y) \equiv y$ 对任意 $(y, x) \in G$. 乘积为: $(z, y)(y, x) = (z, x)$ 对于 $(x, y), (y, z) \in G, x, y, x \in X$. 尤其, 所有元素都相关的等价关系记为 X^2, 称为 X 上的粗糙群胚.

(iii) 设群 H 左作用在非空集合 X 上. 对任意 $h \in H, x \in X$, 记 hx 为元素 h 左作用在 x 上. 令 $G = \{(y, h, x) \mid x, y \in X, h \in H, y = hx\}$. 那么, $G^{(0)} = X \equiv \{(x, e, x) \mid x \in X\}$, 这里 e 为群 H 的单位元, $s(y, h, x) = x$ 和 $t(y, h, x) = y$. 则 $(z, k, y)(y, h, x) = (z, kh, x)$.

例 1.1.6(群胚配对 [11]) 设 G 是一个基 P 上的一个群胚带有源映射 s 和靶映射 t: $s, t : G \longrightarrow P$. 设 $p : E \longrightarrow P$ 是一个映射 (有时称为**纤维丛**(fiber bundle)). 一个 G 在 p 上的**左作用** 是一个映射 $\triangleright : G_t \times_p E \longrightarrow E$ 使得下面条件成立:

$$p(g \triangleright x) = s(g);$$
$$g \triangleright (h \triangleright x) = gh \triangleright x;$$
$$\mathbf{id}\, p(x) \triangleright x = x.$$

对任意 $g, h \in G, x \in E$.

因此, 如果 $E_b = p^{-1}(b)$, 那么 $g \in G$ 的作用是一个同构 $g \triangleright (--) : V_{t(g)} \longrightarrow V_{s(g)}$. 设 $x \in E$ 且定义

$O_x = \{g \triangleright x \mid g \in G, t(g) = p(x)\}$ 称为 x 的**轨道**(orbit);

$G^x = \{g \in G \mid g \triangleright x = x\} = G(p(x))$ 称为 x 的**迷向子群**(isotropy subgroup).

已知 G 在 $p : E \longrightarrow P$ 和 $p' : E' \longrightarrow P$ 上有两个作用, 那么一个映射 $\phi : E \longrightarrow E'$ 称为是**缠绕**(intertwine) 作用如果 $p = p'\phi$ 和 $\phi(g \triangleright x) = g \triangleright \phi(x)$, 对任意 $g \in G, x \in E$ 使得 $t(g) = p(x)$.

一个作用称为**平凡的**如果存在一个集合 X 使得 $E = P \times X$, 且 p 是到第一个分支的投射, 且有 $g \triangleright (t(g), x) = (s(g), x)$, 对任意 $x \in X, g \in G$.

类似地, 一个 G 在 p 上的**右作用** 是一个映射 $\triangleleft : E_p \times_s G \longrightarrow E$ 使得

$$p(x \triangleleft g) = t(g);$$
$$(x \triangleleft g) \triangleleft h = x \triangleleft gh;$$
$$x \triangleleft \mathbf{id}\, p(x) = x$$

对任意 $g, h \in G, x \in E$.

一个群胚配对是在同一个基 P 上的一对群胚 $t, b : V \longrightarrow P$, $l, r : H \longrightarrow P$(这里简用 "top", "bottom", "left" 和 "right") 带有一个 H 在 $t : V \longrightarrow P$ 上的左作用: $\triangleright : H_r \times_t V \longrightarrow V$ 和 V 在 $r : H \longrightarrow P$ 上的右作用: $\triangleleft : H_r \times_t V \longrightarrow H$, 满足下面条件:

$$b(x \triangleright g) = l(x \triangleleft g);$$
$$x \triangleright fg = (x \triangleright f)((x \triangleleft f) \triangleright g);$$
$$xy \triangleleft g = (x \triangleleft (y \triangleright g))(y \triangleleft g)$$

对任意 $f, g \in V, x, y \in H$ 且使得所有可能合成成立.

容易证明

$$x \triangleright \mathbf{id}\, r(x) = \mathbf{id}\, l(x), \quad \forall x \in H;$$
$$\mathbf{id}\, t(g) \triangleleft g = \mathbf{id}\, b(g), \quad \forall g \in V.$$

例 1.1.7(Weyl 群胚 [89]) 设 k 是特征 0 的域, $d \in \mathbb{N}$, 那么集合

$$\widetilde{W} := \{(T, B) \mid T \in \mathrm{Aut}_{\mathbb{Z}}(\mathbb{Z}^d), B\text{ 是 }\mathbb{Z}^d\text{ 的一组基}\}$$

是一个群胚, 带有偏乘法:

$$(T_1, B_1) \circ (T_2, B_2) = \begin{cases} (T_1 T_2, B_2), & \text{如果 } T_2(B_2) = B_1, \\ 0, & \text{其他.} \end{cases}$$

群胚 $\widetilde{W}$ 自然作用在 $\mathbb{Z}^d$ 的所有基集合上, 作用为

$$(T, B_1) \cdot B_2 = \begin{cases} T(B_1), & \text{如果} B_1 = B_2, \\ \text{不定义}, & \text{其他.} \end{cases}$$

设 $\chi : \mathbb{Z}^d \times \mathbb{Z}^d \longrightarrow k^* = k \setminus \{0\}$ 是 $\mathbb{Z}^d$ 上的**双特征**[16]. 即, 满足下面条件:

$$\chi(0, x) = \chi(x, 0) = 1,$$
$$\chi(x + y, z) = \chi(x, z)\chi(y, z),$$
$$\chi(x, y + z) = \chi(x, y)\chi(x, z)$$

对任意 $x, y, z \in \mathbb{Z}^d$.

设 $E = \{e_1, \cdots, e_d\}$ 是 $\mathbb{Z}^d$ 的一组基. 设 $q_{ij} := \chi(e_i, e_j)$ 对任意 $i, j \in \{1, 2, \cdots, d\}$.

对任意 $f, g \in \mathbb{Z}^d$, 定义 $m_{f,g} \in \mathbb{N}_0 \bigcup \{\infty\}$ 如下:

$$m_{f,g} := \min\{m \in \mathbb{N}_0 \mid \chi(f,f)^m \chi(f,g)\chi(g,f) = 1, \text{或} \chi(f,f)^{m+1} = 1, \chi(f,f) \neq 1\}. \tag{1.2}$$

为了方便起见, 如果方程 (1.2) 右边数不存在时, 我们定义 $m_{f,g} = \infty$. 对任意基 $B \in \mathbb{Z}^d$ 和 $f \in B$, 定义 (若可能的话) $s_{f,B} \in \mathrm{Aut}_{\mathbb{Z}}(\mathbb{Z}^d)$ 为

$$s_{f,B}(g) = \begin{cases} -f, & \text{如果} f = g, \\ g + m_{f,g}f, & \text{如果} g \in B, f \neq g. \end{cases}$$

假设 $F \in \mathbb{Z}^d, f \in F, f' \in F \setminus \{f\}$, 那么数 $m(f, f') := m_{f, f'}$ 存在. 设 $m(f, f') = -2$, 那么存在唯一线性映射 $s_{f,F} \in \mathrm{Aut}_{\mathbb{Z}}(\mathbb{Z}^d)$ 使得

$$s_{f,F}(f') = f' + m(f, f')f$$

对任意 $f' \in F$.

显然我们有 $s_{f,B}$ 是良定义当且仅当所有数 $m_{f,f'}$, 对 $f' \in B \setminus \{f\}$ 是有限的.

那么结合双特征 χ 和基 E 的 Weyl 群胚 $W_{\chi,E}$ 是群胚 $\widetilde{W}$ 的最小子群胚, 它满足下面性质:

(1) $(\mathbf{id}, E) \in W_{\chi,E}$.

(2) 如果 $(\mathbf{id}, F) \in W_{\chi,E}$ 成立而且 $s_{f,F}$ 有定义, 那么关系: $(s_{f,F}, F) \in W_{\chi,E}$ 和 $(\mathbf{id}, s_{f,F}(F)) \in W_{\chi,E}$ 也成立.

进一步地, 设 $W_{\chi,E}^{\mathrm{ext}}$ 是群胚 $\widetilde{W}$ 的最小子群胚, 且包含 $W_{\chi,E}$ 和所有对 (T, F) 具有性质 $(\mathbf{id}, F) \in W_{\chi,E}, T(F) = F$. Weyl 群胚 $W_{\chi,E}$ (分别 $W_{\chi,E}^{\mathrm{ext}}$) 称为**全的**, 如果 $s_{f,F}$ 是良定义当 $(\mathbf{id}, F) \in W_{\chi,E}$ (分别 $(\mathbf{id}, F) \in W_{\chi,E}^{\mathrm{ext}}$) 和 $f \in F$.

容易得

$$W^{\text{ext}}_{\chi,E} = \{(\tau T, F) \mid (T, F) \in W_{\chi,E}, \tau \in \text{Aut}_{\mathbb{Z}}(\mathbb{Z}^d), \tau T(F) = T(F)\}.$$

如果 $W_{\chi,E}$ (或等价 $W^{\text{ext}}_{\chi,E}$) 是全的和有限的, 那么三元数据 (Δ, χ, E) 称为是**算术根系**(arithmetic root system), 这里

$$\begin{aligned} \Delta &= \bigcup \{F \mid (\mathbf{id}, F) \in W_{\chi,E}\} \\ &= \bigcup \{s_{f_k, F_{k-1}} \cdots s_{f_2, F_1} s_{f_1, F_0}(E) \mid k \geqslant 0, f_i \in F_{i-1}, \\ & \quad F_0 = E, F_i = s_{f_i, F_{i-1}}(F_{i-1}), 1 \leqslant i \leqslant k\}. \end{aligned}$$

作为定义的一个推论, 有: $\Delta = -\Delta$, 这是因为 $-f = s_{f,F}(f) \in s_{f,F}(F) \subset \Delta$ 对任意 $f \in F$ 具有 $(\mathbf{id}, F) \in W_{\chi,E}$. (注记: Δ 是有限的当且仅当 $W_{\chi,E}$ 是有限的).

定义 1.1.8 相关于嘉当概型 (Cartan scheme) **(也称嘉当图** (Cartan graph)) **的 Weyl 群胚** [50, 51, 89, 95].

设 I 是一个非空有限集合且 $\{\alpha_i \mid i \in I\}$ 是 $\mathbb{Z}^I$ 的一组标准基. 一个**广义嘉当矩阵** $C = (c_{ij})_{i,j \in I}$ 是在 $\mathbb{Z}^{I \times I}$ 中的一个矩阵, 使得下面两个条件成立:

(M1) $c_{ii} = 2$ 且 $c_{jk} \leqslant 0$ 对任意 $i, j, k \in I, j \neq k$;

(M2) 如果 $i, j \in I$ 且 $c_{ij} = 0$, 那么 $c_{ji} = 0$.

设 A 是一个非空集合, $r_i : A \longrightarrow A, i \in I$ 表示一个映射且 $C^a = (c^a_{jk})_{j,k \in I}$ 是在 $\mathbb{Z}^{I \times I}$ 中的一个广义嘉当矩阵, 对任意 $a \in A$. 那么四重数 (quadruple) 组 $\mathcal{C} = \mathcal{C}(I, A, (r_i)_{i \in I}, (C^a)_{a \in A})$ 称为是**嘉当概型**如果下面两个条件成立:

(C1) $r_i^2 = \mathbf{id}$ 对任意 $i \in I$;

(C2) $c^a_{ij} = c^{r_i(a)}_{ij}$ 对任意 $a \in A$ 和 $i, j \in I$.

我们说 $\mathcal{C}$ 是**连通的**如果群 $\langle r_i \mid i \in I \rangle \subset \text{Aut}(A)$ 可迁移 (transitively) 作用在 A 上, 即对 $a, b \in A$ 且 $a \neq b$ 存在 $n \in \mathbb{N}, a_1, a_2, \cdots, a_n \in A$ 和 $i_1, i_2, \cdots, i_{n-1} \in I$ 满足:

$$a_1 = a, a_n = b, a_{j+1} = r_{i_j}(a_j) \quad \text{对任意} \quad j = 1, \cdots, n-1.$$

两个嘉当概型 $\mathcal{C} = \mathcal{C}(I, A, (r_i)_{i \in I}, (C^a)_{a \in A})$ 和 $\mathcal{C}' = \mathcal{C}'(I', A', (r'_i)_{i \in I'}, (C'^a)_{a \in A'})$ 称为是**等价的**, 如果存在双射 $f_0 : I \longrightarrow I'$ 和 $f_1 : A \longrightarrow A'$ 满足:

$$f_1(r_i(a)) = r'_{f_0(i)}, \quad c'^{f_1(a)}_{f_0(i) f_0(j)} = c^a_{ij}$$

对任意 $i, j \in I, a \in A$.

设 $\mathcal{C} = \mathcal{C}(I, A, (r_i)_{i \in I}, (C^a)_{a \in A})$ 是一个嘉当概型. 对任意 $i \in I$ 和 $a \in A$, 定义 $\sigma_i^a \in \text{Aut}(\mathbb{Z}^I)$ 为

$$\sigma^a(\alpha_j) = \alpha_j - c^a_{ij} \alpha_i \quad \text{对任意} j \in I.$$

那么 $\mathcal{C}$ 的 **Weyl 群胚**是范畴 $\mathcal{W}(C)$ 使得对象 $\mathrm{Ob}(\mathcal{W}(C)) = A$ 且态射由映射 $\sigma_i^a \in \mathrm{Hom}(a, r_i(a))$, $i \in I, a \in A$ 生成. 形式上, 对任意 $a, b \in A$, 集合 $\mathrm{Hom}(a, b)$ 由三元数组 (b, f, a) 组成, 这里

$$f = \sigma_{i_n}^{r_{i_{n-1}}\cdots r_{i_1}(a)} \cdots \sigma_{i_2}^{r_{i_1}(a)} \sigma_{i_1}^a$$

和 $b = r_{i_n} \cdots r_{i_2} r_{i_1}(a)$ 对 $n \in \mathbb{N}_0, i_1, \cdots, i_n \in I$. 合成运算由群 $\mathrm{Aut}(\mathbb{Z}^I)$ 的结构诱导出:

$$(a_3, f_2, a_2) \circ (a_2, f_1, a_1) = (a_3, f_2 f_1, a_1)$$

对 $(a_3, f_2, a_2), (a_2, f_1, a_1) \in \mathrm{Hom}(\mathcal{W}(C))$. 有时, 我们也简单写 $(b, f, a) \in \mathrm{Hom}(a, b)$ 为 $f \in \mathrm{Hom}(a, b)$. I 的基数称为 $\mathcal{W}(C)$ 的**秩**.

一个嘉当型 $\mathcal{C}$ 的 Weyl 群胚 $\mathcal{W}(C)$ 是一个群胚. 事实上, 条件 (M1) 可推出 $\sigma_i^a \in \mathrm{Aut}(\mathbb{Z}^I)$ 是一个反射 (reflection) 对任意 $i \in I, a \in A$, 因此由条件 (C1) 和 (C2) 可知 $\sigma_i^a \in \mathrm{Hom}(a, r_i(a))$ 的逆是 $\sigma_i^{r_i(a)} \in \mathrm{Hom}(r_i(a), a)$. 所以, $\mathcal{W}(C)$ 中每个态射是同构.

如果 C 和 C' 是等价的嘉当概型, 那么 $\mathcal{W}(C)$ 和 $\mathcal{W}(C')$ 是同构群胚.

回顾一个群胚 G 是连通的, 如果对每个 $a, b \in \mathrm{Ob}(G)$, 类 $\mathrm{Hom}(a, b)$ 是非空的. 因此, $\mathcal{W}(C)$ 是连通群胚当且仅当 $\mathcal{C}$ 是连通的嘉当概型. 于是, 说 $\mathcal{C}$ 是**连通的**如果对任意 $a, b \in A$, 存在 $w \in \mathrm{Hom}(a, b) \subset \mathrm{Hom}(\mathcal{W}(C))$. 称 $\mathcal{C}$ 是**单连通的**如果 $\mathrm{Hom}(a, a) = \{\mathbf{id}^a\}$. 设 $w \in \mathrm{Hom}(a, b)$, 定义 w 的**长度**为

$$\ell(w) = \min\{m \in \mathbb{N}_0 \mid w = \sigma_{i_1} \cdots \sigma_{i_m} 1_a, i_1, \cdots, i_m \in I\}.$$

在许多情况下, 很自然假设一个嘉当概型满足下面附加条件:

(C3) 若 $a, b \in A$ 且 $(b, \mathbf{id}, a) \in \mathrm{Hom}(a, b)$, 那么 $a = b$.

设 $\mathcal{C}$ 是嘉当图. 对任意 $a \in A$, 令

$$(R^{\mathrm{re}})^a = \{\mathbf{id}^a \sigma_{i_1} \cdots \sigma_{i_k}(\alpha_j) \mid k \in \mathbb{N}_0, i_1, \cdots, i_k, j \in I\} \subseteq \mathbb{Z}^I.$$

集合 $(R^{\mathrm{re}})^a$ 中元素称为在 a 处的**实根**. 对 $(\mathcal{C}, ((R^{\mathrm{re}})^a)_{a \in A})$ 记为 $R^{\mathrm{re}}(\mathcal{C})$. 一个实根 $\alpha \in (R^{\mathrm{re}})^a, a \in A$ 称为正的 (负的) 如果 $\alpha \in \mathbb{N}_0^I (\alpha \in -\mathbb{N}_0^I)$.

一般地, 我们有下面定义.

定义 1.1.9 设 $\mathcal{C} = \mathcal{C}(I, A, (r_i)_{i \in I}, (C^a)_{a \in A})$ 是一个嘉当概型. 对任意 $a \in A$, 设 $R^a \subseteq \mathbb{Z}^I$, 定义

$$m_{i,j}^a = |R^a \cap (\mathbb{N}_0 \alpha_i + \mathbb{N}_0 \alpha_j)|, \quad i, j \in I, a \in A.$$

我们说 $\mathcal{R} = \mathcal{R}(\mathcal{C}, (R^a)_{a \in A})$ 是型 $\mathcal{C}$ 的一个**根系**(root system), 如果下面条件成立:

(R1) $R^a = R^a_+ \cup -R^a_+$, 这里 $R^a_+ = R^a \cap \mathbb{N}_0^I, \forall a \in A$.

(R2) $R^a \cap \mathbb{Z}\alpha_i = \{\alpha_i, -\alpha_i\}$ 对任意 $i \in I, a \in A$.

(R3) $\sigma_i^a(R^a) = R^{r_i(a)}$ 对任意 $i \in I, a \in A$.

(R4) 如果 $i, j \in I, a \in A$ 满足 $i \neq j$ 且 $m_{i,j}^a$ 是有限的, 那么 $(r_i R_j)^{m_{i,j}^a}(a) = a$.

称 $\mathcal{R}$ 是有限的如果对任意 $a \in A$, 集合 R^a 是有限的. 如果 $\mathcal{R}$ 是 $\mathcal{C}$ 型有限根系, 那么 $\mathcal{R} = \mathcal{R}^{\nabla \rceil}$, 从而 $\mathcal{R}$ 也是 $\mathcal{C}$ 型有限根系.

例 1.1.10 (i) 存在许多嘉当矩阵不同的 Weyl 群胚的例子. 最简单的是: 设 $I = \{1, 2\}, A = \{x, y\}, r_2 = \mathbf{id}, r_1(1) = 2, r_1(2) = 1$,

$$C^x = \begin{pmatrix} 2 & -1 \\ -3 & 2 \end{pmatrix}, \quad C^y = \begin{pmatrix} 2 & -1 \\ -4 & 2 \end{pmatrix}.$$

设 $\mathcal{C}$ 是对应嘉当图. 那么存在唯一的 $\mathcal{C}$ 型根系. 正根为

$$R^x_+ = \left\{ \begin{pmatrix} 1 \\ 0 \end{pmatrix}, \begin{pmatrix} 0 \\ 1 \end{pmatrix}, \begin{pmatrix} 1 \\ 1 \end{pmatrix}, \begin{pmatrix} 1 \\ 2 \end{pmatrix}, \begin{pmatrix} 1 \\ 3 \end{pmatrix}, \begin{pmatrix} 2 \\ 3 \end{pmatrix}, \begin{pmatrix} 3 \\ 4 \end{pmatrix}, \begin{pmatrix} 3 \\ 5 \end{pmatrix} \right\},$$

$$R^y_+ = \left\{ \begin{pmatrix} 1 \\ 0 \end{pmatrix}, \begin{pmatrix} 0 \\ 1 \end{pmatrix}, \begin{pmatrix} 1 \\ 1 \end{pmatrix}, \begin{pmatrix} 1 \\ 2 \end{pmatrix}, \begin{pmatrix} 1 \\ 3 \end{pmatrix}, \begin{pmatrix} 1 \\ 4 \end{pmatrix}, \begin{pmatrix} 2 \\ 3 \end{pmatrix}, \begin{pmatrix} 2 \\ 5 \end{pmatrix} \right\}.$$

(ii) 设 $\alpha_1 = (1, 0, 0), \alpha_2 = (0, 1, 0), \alpha_3 = (0, 0, 1)$, 且

$$R^a_+ := \{\alpha_1, \alpha_2, \alpha_3, (0, 1, 1), (0, 1, 2), (1, 0, 1), (1, 1, 1), (1, 1, 2)\}.$$

对任意 $1 \leq i, j \leq 3$, 定义矩阵 $\mathcal{C}$ 中元素 $c_{i,j}$:

$$c_{i,j} := -\max\{k, | k\alpha_i + \alpha_j \in R^a_+\}, \quad c_{i,i} := 2 \ (i \neq j).$$

于是, 广义嘉当矩阵为

$$\mathcal{C}^a = (c_{i,j})_{i,j} = \begin{pmatrix} 2 & 0 & -1 \\ 0 & 2 & -1 \\ -1 & -2 & 2 \end{pmatrix}.$$

它定义的反射:

$$\sigma_i(\alpha_j) = \alpha_j - c_{ij}\alpha_i, \quad j = 1, 2, 3.$$

例如

$$\sigma_1 = \begin{pmatrix} -1 & 0 & 1 \\ 0 & 1 & 0 \\ 0 & 0 & 1 \end{pmatrix}$$

且
$$\sigma_1(R_+^a) = \{-\alpha_1, \alpha_2, (1,0,1), (1,1,1), (2,1,2), \alpha_3, (0,1,1), (1,1,2)\}.$$

$\sigma_1(R_+^a)$ 中元素要么正、要么负. 设 $R^a = R_+^a \cup -R_+^a$, $R^b = \sigma_1(R^a) =: R_+^b \cup -R_+^b$. 于是, 我们可以从 R_+^b 建立一个嘉当矩阵且给出新的反射. 在这个例子中, 我们有下面嘉当图:

$$\begin{pmatrix} 2 & 0 & -1 \\ 0 & 2 & -1 \\ -1 & -1 & 2 \end{pmatrix} \xrightarrow{\sigma_3} \begin{pmatrix} 2 & -1 & -1 \\ -1 & 2 & -1 \\ -1 & -1 & 2 \end{pmatrix} \xrightarrow{\sigma_2} \begin{pmatrix} 2 & -1 & 0 \\ -1 & 2 & -1 \\ 0 & -1 & 2 \end{pmatrix}$$

$$\downarrow \sigma_1 \qquad\qquad\qquad\qquad\qquad\qquad\qquad \downarrow \sigma_1$$

$$\begin{pmatrix} 2 & 0 & -1 \\ 0 & 2 & -1 \\ -1 & -2 & 2 \end{pmatrix} \qquad\qquad\qquad \begin{pmatrix} 2 & -1 & 0 \\ -1 & 2 & -2 \\ 0 & -1 & 2 \end{pmatrix}$$

注记 1.1.11 设 $\mathfrak{B}$ 是有限维对角型 Nichols 代数. 令 R_+ 是 $\mathfrak{B}$ 的 PBW 生成子集合, 那么 $R_+ \cup -R_+$ 是一个有限 Weyl 群胚的根系 [89].

1.2 拟 群

粗落地说, 一个拟群是一个非空集合 Q 带有一个二元运算 $\cdot : Q \times Q \longrightarrow Q$ 使得对任意 $a \in Q$, 它的左乘和右乘算子是双射. 如存在 $e \in Q$(单位元素) 满足 $ea = a = ae$ 对 $a \in Q$, 那么 $(Q, \cdot, e)$ 称为是一个圈 (loop). 因此一个圈是非结合的群的翻版 [78, 100, 149].

定义 1.2.1 一个**拟群** $(Q, \cdot, \backslash, /)$ 是一个非空集合带有三个二元运算 $\cdot$ (称为乘法 (multiplication)), $\backslash$ (称为左除 (left division)) 和 $/$ (称为右除 (right division)) 使得下面等式成立:

$$a \backslash (ab) = b, \quad a(a \backslash b) = b, \quad (ab)/b = a\,(a/b)b = a$$

对任意 $a, b \in Q$. 如果 $a\, a = b/b$ 对任意 $a, b \in Q$, 那么我们称拟群为**圈**(loop).

尤其, 一个可逆性拟群或 IP 圈是一个非空集合带有一个乘积、单位元 e 和下面性质: 对任意 $u \in G$, 存在 $u^{-1} \in G$ 使得

$$u^{-1}(uv) = v, \quad (vu)u^{-1} = v, \quad \forall v \in G.$$

1.2 拟 群

一个拟群称为是**灵活的**(flexible), 如果 $u(vu) = (uv)u, \forall u, v \in G$; 进一步, 称为是**交错的**(alternative), 如果也有 $u(uv) = (uu)v, u(vv) = (uv)v$ 对任意 $u, v \in G$. 拟群称为是**魔方**(moufang), 如果 $u(v(uw)) = ((uv)u)w$ 对任意 $u, v, w \in G$.

注记 1.2.2 在任何一个拟群 G 中, 任意一个元素都有唯一的逆元, 且 $(u^{-1})^{-1} = u$, $(uv)^{-1} = v^{-1}u^{-1}$ 对任意 $u, v \in G$.

下面列出拟群的一些性质, 证明见文献 [100].

命题 1.2.3 (1) 设 G 是一个拟群. 那么下面等式等价: 对任意 $u, v, w \in G$:

(i) $u(v(uw)) = ((uv)u)w$.

(ii) $((uv)w)v = u(v(wv))$.

(iii) $(uv)(wu) = (u(vw))u$.

(2) 设 G 是灵活的拟群. 那么我们有: $u(vu^{-1}) = (uv)u^{-1}$ 对任意 $u, v \in G$.

设 G 是一个拟群. 我们引进一个可乘结合子 $\phi : G^3 \longrightarrow G$ 为

$$(uv)w = \phi(u, v, w)u(vw)$$

对任意 $u, v, w \in G$. 由此, 可推出

$$\phi(u, v, w) = ((uv)w)((w^{-1}v^{-1})u^{-1}).$$

现在定义一个可结合元素或中心 (nucleus) $N(G)$ 群如下:

$$N(G) = \{a \in G \mid (au)v = a(uv), u(av) = (ua)v, (uv)a = u(va), \forall u, v \in G\}.$$

容易验证 $N(G)$ 是一个群. 我们说一个拟群是拟结合的, 如果 ϕ 和所有它的共轭 $u\phi u^{-1}$ 的像在 $N(G)$ 中, 且一个拟群是中心的, 如果 ϕ 在中心 $Z(G)$.

命题 1.2.4 设 G 是一个拟群带有单位元 e 和可乘结合子 ϕ. 那么

(i) $N(G) = \{a \in G \mid \phi(a, u, v) = \phi(u, a, v) = \phi(u, v, a) = e, \forall u, v \in G\}$.

(ii) $\phi(e, u, v) = \phi(u, e, v) = \phi(u, v, e) = \phi(u, u^{-1}, v) = \phi(u, v, v^{-1}) = \phi(uv, v^{-1}, u^{-1}) = \phi(u^{-1}, uv, v^{-1}) = \phi(v^{-1}, u^{-1}, uv) = e, \forall u, v \in G$.

(iii) $\phi(au, v, w) = a\phi(u, v, w)a^{-1}, \phi(ua, v, w) = \phi(u, av, w), \phi(u, va, w) = \phi(u, aw), \phi(u, v, wa) = \phi(u, v, w), \forall u, v, w \in G$ 和 $a \in N(G)$.

(iv) 如果 G 是拟结合的, 则 ϕ 是一个伴随 3-余循环, 即满足:

$$\phi(u, v, w)\phi(u, vw, z)(u\phi(v, w, z)u^{-1}) = \phi(uv, w, z)\phi(u, v, wz)$$

对任意 $u, v, w, z \in G$.

(v) G 是灵活的当且仅当 $\phi(u, v, u) = e$ 对任意 $u, v \in G$, 且 G 是交错的如果也有 $\phi(u, u, v) = \phi(u, v, v) = e$ 对任意 $u, v \in G$.

(vi) 一个拟结合拟群 G 是一个魔方当且仅当它是灵活的且下面等式之一成立:

$$u\phi(v,u,w)u^{-1} = \phi(u,vu,w)^{-1}, \quad \phi(u,vw,v) = \phi(u,v,w)^{-1},$$
$$\phi(uv,w,u) = \phi(u,v,w)$$

对任意 $u,v,w \in G$.

1.3 箭 图

箭图是构造 Hopf 代数的主要工具之一 [42, 163, 206].

定义 1.3.1 一个**箭图**是一组 (A,P,s,t), 这里 A 与 P 是非空集合, $s,t: A \longrightarrow P$ 是两个映射, A 中的元素是从它的源 $s(a)$ 到靶 $t(a)$ 的一个"箭", 也称 A 是 P 上的一个箭图. 箭图也称为**有向图** (oriented graph).

对于 $p,q \in P$, 记从 p 到 q 的箭集合为 $A(p,q)$, 尤其, $A(p) = A(p,p)$. 相同 P 上的两个箭图之间的态射按常规方式定义. 在方向上箭图 B 区别于箭图 A 如果 B 与 A 箭相同, 但对于一些箭的源和靶不同. 例如: **反箭图** $A^{\mathrm{op}} = A \times \{-1\}$. 如果 $x \in A$, 那么 $x^{-1} := (x,-1)$ 的源 $s(x^{-1}) = t(x), t(x^{-1}) = s(x)$. 我们也有 $(a^{\mathrm{op}})^{\mathrm{op}} = A$, $(x^{-1})^{-1} = x$ 对任意 $x \in A$.

如果 A 和 B 都是 P 上箭图, 那么非交并 $A \coprod B$ 也是 P 上的一个箭图. A 的偶是 $DA := A \coprod A^{\mathrm{op}}$, 它不依赖于 A 的方向. 有时也记 $\overline{A} = DA$.

设 n 是正整数. 在 A 中, 长度为 n 的**路** (path) 是 A 中元素的一个序列 $w = (x_1, \cdots, x_n)$ 使得 $s(x_{i+1}) = t(x_i)$ 对任意 $1 \leqslant i \leqslant n$. 我们将记它为 $w = x_1 \cdots x_n$. 长度为 0 的路是一个符号 $\mathbf{id}p$, 对任意 $p \in P$. A 中长度为 n 的所有路集合是一个箭图 $\mathrm{Path}_n(A)$ 具有性质 $s(w) = s(x_1), t(w) = t(x_n)$ 如果 $w = x_1 \cdots x_n$ (如果 $n > 0$), 且 $s(\mathbf{id}p) = t(\mathbf{id}p) = p$ (如果 $n = 0$). A 中所有路的箭图是 $\mathrm{Path}(A) = \coprod_{n \geqslant 0} \mathrm{Path}_n(A)$.

箭图 A 诱导 P 上的一个等价关系: $p \approx q$ 当且仅当存在 $w \in \mathrm{Path}(DA)$ 具有 $s(w) = p, t(w) = q$, 对任意 $p,q \in P$. 我们称 A 是连通的如果对任意 $p,q \in P$ 有 $p \approx q$.

定义 1.3.2 一个映射 $f: L \longrightarrow P$ 可以被看成是一个箭图具有 $s = t = f$, 如此的一个箭图称为是**圈丛** (loop bundle). 有时, 也用**纤维** (fiber) 符号 $L_p := L(p) = f^{-1}(p)$.

例 1.3.3 一个箭图 A 可以给出两个圈丛: $s: A \longrightarrow P$ 和 $t: A \longrightarrow P$, 分别记为 A^s 和 A^t. 有

$$A^s = \coprod_{p \in P} A^s(p), \quad A^s(p) := \{(x, x^{-1}) \mid x \in A(p,q), q \in P\}$$

$$A^t = \coprod_{q\in P} A^s(q), \quad A^s(q) := \{(y, y^{-1}) \mid y \in A(p,q), p \in P\}.$$

我们有经典映射 $A \longrightarrow A^t, x \mapsto \bar{x} = (x^{-1}, x)$.

如果 $T: A \longrightarrow B$ 是箭图同态, 那么定义一个圈丛同态 $\overline{T}: A^t \longrightarrow B^t$ 为 $\overline{T}(\bar{x}) = \overline{T(x)}$.

如果 A, B 是 P 上两个箭图, 那么 $A\,{}_t\!\times_s B := \{(a,b) \in A\times B \mid t(a)=s(b)\}$ 是 P 上的一个箭图具有 $s(a,b)=s(a), t(a,b)=t(b)$. 因此, P 上的箭图范畴 $\mathrm{Quiv}(P)$ 是张量范畴具有 $\otimes = {}_t\!\times_s$, 并且具有单位对象 $(P, P, \mathbf{id}, \mathbf{id})$. 但这个张量范畴似乎不是对称的. 然而, 我们有两个自然同构起着弱对称作用. 第一个自然同构:

$$\sigma: B^{\mathrm{op}}\,{}_t\!\times_s A^{\mathrm{op}} \longrightarrow (A\,{}_t\!\times_s B)^{\mathrm{op}}, \quad (y^{-1}, x^{-1}) \mapsto (x,y), \quad 对任意 (x,y) \in A\,{}_t\!\times_s B.$$

第二种可能性是非常相似的. 设 $B\,{}_s\!\times_t A := \{(b,a) \in B\times A \mid s(b)=t(a)\}$ 是 P 上的一个箭图具有 $s(b,a)=s(a), t(b,a)=t(b)$. 那么我们定义

$$\tau: A\,{}_t\!\times_s B \longrightarrow B\,{}_s\!\times_t A, \quad (x,y) \mapsto (y,x), \quad 对任意 (x,y) \in A\,{}_t\!\times_s B.$$

它们的关系如下: 定义 $\mu: B\,{}_s\!\times_t A \longrightarrow (B^{\mathrm{op}}\,{}_t\!\times_s A^{\mathrm{op}})^{\mathrm{op}}$ 为 $\mu(y,x)=(y^{-1}, x^{-1})$. 那么下面图交换:

$$\begin{array}{ccc} A\,{}_t\!\times_s B & \stackrel{\tau}{\longrightarrow} & B\,{}_s\!\times_t A \\ {\scriptstyle \sigma}\nwarrow & & \swarrow {\scriptstyle \mu} \\ & (B^{\mathrm{op}}\,{}_t\!\times_s A^{\mathrm{op}})^{\mathrm{op}} & \end{array}$$

1.4 乘　　子

乘子 (multiplier) 代数的概念可追溯到 G. Hochschild 的上同调与扩张 [90] 以及 B.E.Johnson 的拓扑代数中心化子的工作 [93] 中, 而乘子代数概念的纯代数研究归功于 J. Dauns 在 1969 年的乘子环的工作 [52].

我们总是假设 A 是域上结合代数 (可以没有单位元), 要求乘积是非退化的 (non-degenerate). 也假设 X 有非退化的 A-模作用, 例如: 如果 X 是左 A-模, 当 $ax=0$ 时, 就有 $x=0$ 对 $x \in X$, 对任意 $a \in A$. 有时, 假设作用是单位的, 例如: 对左 A-模 X, 有 $AX=X$. 一般地, 任意单位作用是非退化的, 反之, 不一定真.

定义 1.4.1 设 X 是左 A-模. 记 $Y=\{\rho: A \longrightarrow X \mid \rho(ab)=a\rho(b), a,b \in A\}$.

例如：对任意 $x \in X$，有 $\rho_x \in Y$，定义为：$\rho_x(a) = ax$. 于是，我们有单射：$X \longrightarrow Y, x \mapsto \rho_x$. 事实上，如果 $\rho_x = 0$，那么 $ax = 0$ 对任意 $a \in A$，由于模作用是非退化的，于是有 $x = o$.

命题 1.4.2 Y 是一个非退化的左 A-模，模作用为 $(a\rho)(b) = \rho(ab)$ 对任意 $a, b \in A, \rho \in Y$, X 是 Y 的一个子模.

注记 1.4.3 (i) 如果 A 有单位元，那么 $Y = X$. 因为当 $\rho \in Y, x = \rho(1)$，那么有 $\rho(a) = a\rho(1) = ax = \rho_x(a)$.

(ii) 我们总有 $AY \subseteq X$. 如果左 A-模 X 是单位的，那么有 $X \subseteq AY \subseteq X$，所以 $AY = X$. 然而一般地，AY 是 X 的真子模.

(iii) 对 $y \in Y, a \in A$，一般记 $ay = y(a)$. 有时用 $\overline{X}$ 表示扩张模 Y，称为是 X 的完备化 (completion).

定义 1.4.4 设 X 是非退化的双 A-模. 我们记 $Z = \{(\lambda, \rho) \mid \lambda, \rho : A \longrightarrow X, a\lambda(b) = \rho(a)b, a, b \in A\}$.

注记 1.4.5 (i) 当有如此对 (λ, ρ)，对任意 $a, b, c \in A$ 有

$$\rho(ab)c = ab\lambda(c) = a\rho(b)c.$$

因为 X 有非退化右 A-模作用，所以有 $\rho(ab) = a\rho(b)$. 类似地，$\lambda(ab) = \lambda(a)b$.

(ii) 对任意 $x \in X$，定义 $(\lambda_x, \rho_x) \in Z$ 为

$$\lambda_x(a) = xa, \quad \rho_x(a) = ax \quad \text{对任意 } a \in A.$$

进一步，映射 $X \longrightarrow Z, x \mapsto (\lambda_x, \rho_x)$ 是一个单射.

(iii) 一般地，$AZA \subseteq X$. 如果 A 有单位元，那么 $Z = X$.

命题 1.4.6 对任意 $z = (\lambda, \rho) \in Z, a \in A$. 定义：

$$az = (a\lambda(\cdot), \rho(\cdot a))\ za = (\lambda(a\cdot), \rho(\cdot)a).$$

那么 Z 是 (A, A)-双模，且 X 在 Z 中的嵌入是一个双模映射.

例 1.4.7 (i) 设 A 是一个代数，A 可以被看成 (A, A)-双模，那么我们就有乘子代数 $M(A)$.

(ii) 设 A 是一个代数，$A \otimes A$ 可以被看成 (A, A)-双模，结构为

$$a(b \otimes c) = ab \otimes c, \quad (a \otimes b)c = a \otimes bc, \quad \text{对任意 } a, b, c \in A.$$

那么，这就形成一个子代数：$\{x \in M(A \otimes A) \mid (a \otimes 1)x, x(1 \otimes a) \in A \otimes A, a \in A\}$.

(iii) 设 A 是一个代数，V 是一个向量空间. 设 $X = A \otimes V$，其形成 (A, A)-双模：

$$a(b \otimes v) = ab \otimes v, \quad (a \otimes v)b = ab \otimes v \quad \text{对任意 } a, b \in A, v \in V.$$

命题 1.4.8 设 A,B 是非退化代数,且 $\gamma : A \longrightarrow M(B)$ 是代数同态. 假设存在一个幂等元 $e \in M(B)$ 使得
$$\gamma(A)B = eB, \quad B\gamma(A) = Be.$$
那么存在唯一的 γ 扩张代数同态 $\gamma_1 : M(A) \longrightarrow M(B)$ 使得 $\gamma_1(1) = e$.

注记 如果 $e = 1$,那么命题中的条件变为 γ 是满射 (见文献 [193, 194, 209]).

1.5 环扩张

设 A 是域 k 上的结合的非交换的具有单位元的环, $B \subseteq A$ 是一个子环, 记该环扩张为 $A|B$. A 自然有一个 (A,B)-双模、(B,A)-双模结构和 (B,B)-双模结构. 自然有自同态代数扩张 $\mathrm{End}(A_B)/A$, 即 $A \longrightarrow \mathrm{End}(A_B), a \mapsto \lambda_a$, 这里 λ_a 是左乘法函子, 且是右 B-模自同态.

本节可参考文献 [75, 85, 99, 102, 143—145, 153, 175, 176, 249].

我们记双模 $_BP_B$ 的中心化子为: $P^B := \{p \in P \mid pb = bp \; \forall b \in B\}$. 作为特殊例子, B 在 A 中的中心化子代数为: $A^B = C_A(B)$. 代数扩张 $A|B$ 称为**不可约的**(irreducible), 如果中心化子代数是平凡的, 即 $C_A(B) = k1$. 因为 $Z(A), Z(B) \subseteq C_A(B)$, 所以它们也是平凡的.

设 $\Lambda := \mathrm{End}(A_B)$, 那么
$$C_\Lambda(A) = \{f \in \Lambda \mid af(x) = f(ax), \forall a \in A\} = \mathrm{End}(_AA_B) \cong C_A(B)^{\mathrm{op}},$$
$T := (A \otimes_B A)^B$ 有代数结构, 来自于 $T \cong \mathrm{End}(_A(A \otimes_B A)_A)$ 如下:
$$tt' = t'^1 t^1 \otimes t^2 t'^2, \quad 1_T = 1 \otimes 1,$$
这里 $t = t^1 \otimes t^2 \in T$.

我们习惯说扩张 $A|B$ 有左或右性质 X(例如有限投射), 如果自然模 $_BA$ 或 A_B 有性质 X.

代数扩张 $A|B$ 称为**可裂扩张**(split extension), 如果存在一个 B-双模投射 $E : A \longrightarrow B$. 因此, $E(1) = 1$, $E(xay) = xE(a)y$ 对任意 $a \in A, x,y \in B$, 且 $A = B \oplus \mathrm{Ker}(E)$ 作为 B-双模. 代数扩张 $A|B$ 称为 **Frobenius 扩张**, 如果自然的右 B-模 A_B 是有限生成投射且存在 M 与它的代数扩张对偶的双模同构: $_BA_A \cong {_B\mathrm{Hom}}(A_B, B_B)_A$. 这等价于代数扩张 $A|B$ 有一个双模同态 $E : {_BA_B} \cong {_BB_B}$(称为是 Frobenius 同态) 和存在 A 中元素 $\{x_i\}_i^n, \{y_i\}_i^n$ (称为对偶基), 使得下式成立:
$$\sum_{i=1}^n E(ax_i)y_i = a = \sum_{i=1}^n x_i E(y_i a)$$

对任意 $a \in A$. 进一步, 代数扩张 $A|B$ 是可裂的, 当且仅当存在元素 $d \in C_A(B)$ 使得 $E(d) = 1$.

如果 A_B 是**自由的**(free), 则称代数扩张 $A|B$ 是自由 Frobenius 扩张. 可选择 A_B 的对偶基 $\{x_i\}, \{f_i\}$, 满足 $f_i(x_j) = \delta_{ij}$. 可得到一组正交对偶基 $\{x_i\}, \{y_i\}$ 使得 $E(y_i x_j) = \delta_{ij}$. 反之, 具有 $E, \{x_i\}, \{y_i\}$ 满足这个方程的扩张 $A|B$ 是自由 Frobenius.

尤其, $B = k1$, 那么 A 是 Frobenius 代数 (这个概念首次出现在 1903 年 Frobenius 的文章中), 这等价于存在一个忠实的 (或非退化的) 线性函数 $E: A \longrightarrow k$, 即 $E(Aa) = 0$ 可推出 $a = 0$, 或等价地, $E(aA) = 0$ 可推出 $a = 0$.

命题 1.5.1　如果 $A|B$ 为 Frobenius 扩张带有 Frobenius 同态 $E: A \longrightarrow B$ 及对偶基 $\{x_i, y_i\}$, 且 $R|A$ 为 Frobenius 扩张带有 Frobenius 同态 $F: R \longrightarrow A$ 及对偶基 $\{z_j, w_j\}$. 那么 $R|B$ 为 Frobenius 扩张带有 Frobenius 同态 $E \circ F: R \longrightarrow B$ 及对偶基 $\{z_j x_i, y_i w_j\}$.

如果 $A|B$ 和 $R|A$ 是不可约的, 那么合成指数满足拉格朗日 (Lagrange) 方程:

$$[R:B]_{EF} = [R:A]_F [A:B]_E.$$

已知一个 Frobenius 同态 $E: A \longrightarrow B$ 和元素 $c \in C_A(B)$. 定义映射 B-双模映射 cE, Ec 为: $cE(x) := E(xc)$ 和 $Ec(x) := E(cx)$, 这属于 B-双模同态集合 $\mathrm{Hom}_{-B}(A_B, B_B)$ 和 $\mathrm{Hom}_{B-}({}_B A, {}_B B)$ 的 B-中心化子. 因为我们有双模同构: ${}_B A_A \cong {}_B \mathrm{Hom}_{-B}(A, B)_A, a \mapsto Ea$. 于是, 存在唯一的元素 $c' \in C_A(B) = A^B$ 使得 $Ec' = cE$. 那么, 在 $C_A(B)$ 上的映射 $q: c \mapsto c'$ 显然是自同构, 称为是 **Nakayama 自同构**或**模自同构**(modular automorphism)[234], 也可定义为

$$E(q(c)a) = E(ac)$$

对任意 $c \in C_A(B), a \in A$.

$A|B$ 是一个对称 Frobenius 扩张, 如果 q 是一个内自同构. 当 $B = k1$ 时, 正好是通常的对称代数的概念 (即一个具有非退化或忠实的迹的有限维代数), 因为如果 $q(a) = uau^{-1}$, 由上面公式可得, Eu 是一个迹.

我们将考虑 $A \otimes_B A$ 具有 (A, A)-双模结构. $A|B$ 称为是**可分扩张**(separable extension), 如果乘法满射 $m: A \otimes_B A \longrightarrow A$ 作为 (A, A)-双模同态有右逆. 显然, 这等价于: 存在一个元素 $e \in A \otimes_B A$ 使得 $ae = ea$ 和 $m(e) = 1$, 对任意 $a \in A$. 该元素称为是**可分元素**(separability element).

可分扩张实际上是一个具有平凡的相对 Hochschild 上同调群次数是 1 或更多的代数扩张. 一个 Frobenius 扩张 $(A|B, E, x_i, y_i)$ 是可分的 (separable) 当且仅当存在一个元素 $d \in C_A(B)$ 使得 $\sum_i x_i d y_i = 1$. 如果 $B = k1$, 那么 $A|B$ 是可分扩张当且仅当 A 是 k-可分代数, 即: 一个有限维的、半单的 k-代数, 其中在可除代数 D_i

1.5 环扩张

上具有矩阵块 (matrix blocks), 这里 $Z(D_i)$ 是 k 上的有限可分域扩张. 如果 k 是代数闭域, 那么每个 $D_i = k$ 且 A 同构于 k 上 n_i 阶矩阵块的直积.

例 1.5.2 如果 $E'|F'$ 是一个有限可分域扩张, $\alpha \in E'$ 是一个本原元使得 $E' = F'(\alpha)$, 且 $p(x) = x^n - \sum_{i=0}^{n-1} c_i x^i$ 是 α 在 $F'[X]$ 上的极小多项式. 那么可分元为

$$\sum_{i=0}^{n-1} \alpha^i \otimes_{F'} \frac{\sum_{j=0}^{i} c_j \alpha^j}{p'(\alpha)\alpha^{i+1}}.$$

一个 k-代数 A 被称为是 Kanzaki 意义下的强可分的, 如果 A 有一个对称可分元素 e(必然是唯一的), 即: $\tau(e) = e$, 这里 τ 是 $A \otimes_k A$ 上的扭曲映射. 一个等价条件是 A 有一个迹 $t: A \longrightarrow k$(即: $t(ab) = t(ba)$ 对任意 $a, b \in A$), 且存在元素 $x_1, \cdots, x_n, y_i, \cdots, y_n$ 使得 $\sum_i t(ax_i)y_i = a$ 对任意 $a \in A$ 和 $\sum_i x_i y_i = 1_A$. 这推出 k 的特征不能整除矩阵块的阶数 n_i (即 $n_i 1_k \neq 0$).

定义 1.5.3 一个 k-代数扩张 $A|B$ 称为是一个强可分、不可约扩张, 如果 $A|B$ 是一个不可约 Frobenius 扩张带有 Frobenius 同态 $E: A \longrightarrow B$ 和对偶基 $\{x_i\}, \{y_i\}$ 满足条件: ① $E(1) \neq 0$; ② $\sum_i x_i y_i \neq 0$.

注记 因为 $A|B$ 是不可约的, 那么 A 和 B 是平凡的, 所以存在一个非零数 $\mu \in k$ 使得 $E(1) = \mu 1_S$. 那么 $\frac{1}{\mu}E, \mu x_i, y_i$ 是 $A|B$ 的一组新的 Frobenius 同态体系. 不失一般性, 假设

$$E(1) = 1.$$

由此, 作为 (B, B)-双模, 可推出 $A = B \bigoplus \mathrm{Ker}E$ 且 $E^2 = E$, 这里 E 被作为 $\mathrm{End}_B(A)$ 中元素. 同时, 存在非零数 $\lambda \in k$, 有

$$\sum_i x_i y_i = \lambda^{-1} 1_A.$$

于是, $\xi \sum_i x_i \otimes y_i$ 是可分元素, 则 $A|B$ 是可分的. 对于强可分、不可约扩张, 数组 E, x_i, y_i 是唯一决定的.

注记 1.5.4 在 Kanzaki 强可分 k-代数与强可分扩张 $A|k1$ 之间存在一个密切而又复杂的关系. 例如: 设 F_2 是特征 2 的域, $A = M_2(F_2)$ 不是 Kanzaki 强可分的, 但是一个强可分扩张 $A|F_2 1$. 这是因为 $E(A) = a_{11} + a_{12} + a_{21}$, 且

$$\sum_i x_i \otimes y_i = e_{11} \otimes e_{21} + e_{12} \otimes e_{11} + e_{12} \otimes e_{21} + e_{22} \otimes e_{12} + e_{22} \otimes e_{22} + e_{21} \otimes e_{22}$$

满足 $\sum_i x_i y_i = 1$, $E(1) = 1$, E 是 Frobenius 同态具有对偶基 x_i, y_i.

然而, 一个具有 Markov 迹的强可分扩张 $A|k1$ 是 Kanzaki 强可分的. 反之, 设 $k = Z(A)$.

现在回忆一下同态环定理.

定理 1.5.5 $\Lambda|A$ 是一个强可分、不可约扩张, 指数为 λ^{-1}.

证明 对 Frobenius 扩张 $A|B$, 有 $\Lambda \cong A \otimes_B A, f \mapsto \sum_i f(x_i) y_i$, 逆映射为 $a \otimes b \mapsto \lambda_a E \lambda_b$. 记 $A_1 = A \otimes_B A$, 它的 E-乘法为

$$(a \otimes b)(x \otimes y) = aE(bx) \otimes y$$

对任意 $a, b, x, y \in A$. 单位元 $1_1 := \sum_i x_i \otimes y_i$. 容易看出 $E_A := \lambda \mu$(这里 λ 是乘法映射: $A_1 \longrightarrow A$) 是正则 Frobenius 同态, $\{\lambda^{-1} x_i \otimes 1\}, \{1 \otimes y_i\}$ 是对偶基. □

现在描述第一个 Jones 幂等元 $e_1 := 1 \otimes 1 \in A_1$, 作为 (A, A)-双模, 它循环生成 A_1: $A_1 = \{\sum_i x_i e_1 y_i \mid x_i, y_i \in A\}$. 有时, 也称满足 $E(1) = 1$ 的 Frobenius 同态 E 为**条件期望**(conditional expectation). 称 A_1, e_1, E_A 为 $B \subseteq A$ 的**基本结构**.

由基本结构, 对于 $B \subseteq A$, 可以构造**Jones 塔**(Jones tower) 如下:

$$B \subseteq A \subseteq A_1 \subseteq A_2 \subseteq \cdots,$$

这里 A_2 是 $A \subseteq A_1$ 的基本结构, 等等. 即, $A_2 = A_1 \otimes_A A_1 \cong A \otimes_B A \otimes_B A$ 带有 E_A-乘法, 条件期望为: $E_{A_1} = \lambda \mu : A_2 \longrightarrow A_1$, 定义为

$$a \otimes b \otimes c = \lambda a E(b) \otimes c.$$

第二个 Jones 幂等元为: $e_2 = 1_1 \otimes 1_1 = \sum_{i,j} x_i \otimes y_i x_j \otimes y_j$ 且满足 $e_2^2 = e_2$ 在 A_2 的 E_A-乘法中.

注意, $1_2 = \sum_i \lambda^{-1} x_i \otimes 1 \otimes y_i$, $E_{A_i}(e_{i+1}) = \lambda 1$, 这里 $A_0 = A$. 那么在 A_2 中, 关于 e_1, e_2 的下面关系容易得到 (没有不可约假设).

命题 1.5.6 我们有

$$e_1 e_2 e_1 = \lambda e_1 1_2, \quad e_2 e_1 e_2 = \lambda e_2.$$

定义 1.5.7 (1) 一个环扩张 $A|B$ 是左**深度 2**(D2) 扩张, 如果张量平方 $A \otimes_B A$, 作为自然 (B, A)-双模, 同构 A 与自身的有限直和的直和项. 等价地, $A|B$ 是左深度 2 拟基 (quasibase), 如果存在 $\beta_i \in S := \text{End}(_B A_B), t_i \in T := (A \otimes_B A)^B$ 使得, $a, a' \in A$

$$a \otimes a' = \sum_{i=1}^{n} t_i \beta_i(a) a'.$$

(2) 一个环扩张 $A|B$ 是右深度 2 扩张, 如果张量平方 $A \otimes_B A$, 作为自然 (A, B)-双模, 同构 A 与自身的有限直和的直和项. 等价地, $A|B$ 是右深度 2 拟基 (quasi-base), 如果存在 $\gamma_j \in S := \mathrm{End}(_B A_B)$, $u_j \in T := (A \otimes_B A)^B$ 使得, $a, a' \in A$

$$a \otimes a' = \sum_{j=1}^{n} a\gamma_j(a')u_j.$$

(3) 一个环扩张 $A|B$ 是深度 2(D2) 扩张, 如果环扩张 $A|B$ 既是左深度 2 扩张又是右深度 2 扩张. 等价地, 环扩张 $A|B$ 既是左深度 2 拟基又是右深度 2 拟基.

命题 1.5.8 环扩张 $A|B$ 是右深度 2 扩张当且仅当它的反代数扩张 $A^{\mathrm{op}}|B^{\mathrm{op}}$ 是左深度 2 扩张. 进一步, 有经典代数同构: $S(A|B) \cong S(A^{\mathrm{op}}|B^{\mathrm{op}})$ 和 $T(A|B) \cong T(A^{\mathrm{op}}|B^{\mathrm{op}})$.

下面介绍 1933 年由 Ore 引进的扩张 (**Ore 扩张**). 许多 Hopf 代数的例子可以由此扩张得到 [17, 148].

定义 1.5.9 设 A 是一个环具有一个自同构 σ. 设 δ 是一个 σ-导子, 即满足:

$$\delta(ab) = \sigma(a)\delta(b) + \delta(a)b$$

对任意 $a, b \in A$. 那么, 结合参数 $\{\sigma, \delta\}$ 的 A 的 Ore 扩张是一个环 B, 包含 A 为子环, B 是由 A 中元素与一个新变量 y 生成且满足下面等式:

$$ya = \sigma(a)y + \delta(a)$$

对任意 $a \in A$. 记: $B = A[y; \sigma, \delta]$. 一般地, 有

$$(ax^m)(bx^n) = \sum_{i=0}^{m} a\pi_i^m(b)x^{i+n}$$

对任意 $a, b \in B, m, n \in \mathbb{N}$, 这里 π_i^m 表示 i 个 σ 所有 $\binom{m}{i}$ 个可能的合成的和.

部分对偶地, 我们有如下定义.

定义 1.5.10 设 C 是一个余代数具有一个自同构 σ. 设 δ 是一个 σ-**余导子**, 即满足:

$$\Delta(\delta(c)) = \sum \sigma(c_1) \otimes \delta(c_2) + \sum \delta(c_1) \otimes c_2$$

对任意 $c \in C$.

1.6 辫子与纽结

经典纽结理论是处理图 (diagram) 与不变量 (invariant) 问题. 本节主要给出纽结和链接 (link) 概念的简单介绍. 这部分与文献中有向 (余) 代数[158-160] 和

Hopf(余) 代数的理论相关 (参考文献: [49, 98, 121, 142, 157, 162, 184]), 最后我们也介绍了辫子向量以及由此定义的 Nichols 代数 [10, 13].

定义 1.6.1　一个**纽结**如图 1 所示 (它可能是非常复杂的).

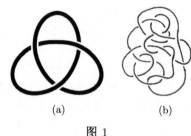

图 1

一般关心的一个问题是: 给出一个纽结的两个平面图, 我们是否能够辨别它们代表的是同一个纽结? 或者, 给出一个辫子的平面图, 如何确定它是否可以被解开结? 这不是一个简单的问题如图 1(b) 所示的纽结. 为此, 我们需要引进如下概念.

定义 1.6.2　给定一个平面上的两个 n 点集, 第一个集合的点通过一股绳子 (strand) 与第二个集合中的点连接 (在空间上). 这样的连接称为**辫子**. 辫子可以如图 2 合成.

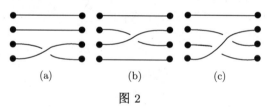

图 2

注记 1.6.3　绳子上的连接与移动有如下规则:

(i) 绳子从左往右移动.

(ii) 绳子可以被拉伸, 然后可以被认为是一样的.

(iii) 绳子之间可以上下移动.

(iv) 有时, 绳子需要从一个到另一个绳子的上面或下面穿过. 我们通常区分出下图的不同点.

(v) 图 3 和图 4 中的两种连接, 可以通过拉直绳子, 变成同样的辫子图.

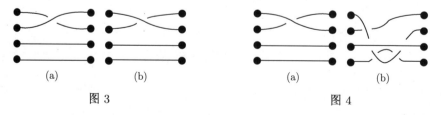

图 3　　　　　　　　图 4

注记 1.6.4 所有的绳子必须从左向右移动. 图 5 中的纽结就不能看成是一个辫子.

接下来, 我们通过下面的一个命题来给出辫子群 B_4 的定义.

定义 1.6.5 继续考虑像上面的那些带有 4 根绳子的辫子. 我们将下面的辫子分别表示为 $\sigma_1, \sigma_2, \sigma_3$, B_4 中的每一个辫子都可以由 $\sigma_1, \sigma_2, \sigma_3$ 或它们的逆合成得到. 换句话说, $\sigma_1, \sigma_2, \sigma_3$ 生成了 B_4. 进一步地, $\sigma_1, \sigma_2, \sigma_3$ 满足下面的关系:

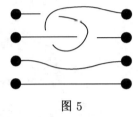

图 5

$$\sigma_1\sigma_2\sigma_1 = \sigma_2\sigma_1\sigma_2, \quad \sigma_2\sigma_3\sigma_2 = \sigma_3\sigma_2\sigma_3.$$

同样地, 也有 $\sigma_1\sigma_3 = \sigma_3\sigma_1$.

现在继续解决上面提出的问题. 我们的想法是:

(i) 通过闭合一个辫子来得到一个纽结 (或链接);

(ii) 利用 Temperley-Lieb 代数 $\mathrm{TL}_n(q)$ 上一个特殊函数.

设 $q \in \mathbb{C}$. Temperley-Lieb 代数 $\mathrm{TL}_n(q)$ 是一个结合 $\mathbb{C}$-代数, 具有生成子 $f_1, f_2, \cdots, f_n$ 满足下面关系式:

$$f_i^2 = -(q + q^{-1})f_i \quad \forall i;$$
$$f_i f_{i\pm 1} f_i = f_i \quad \forall i;$$
$$f_i f_j = f_j f_i, \quad 如果 |i - j| \geqslant 2.$$

(iii) 联合这个函数来得到 q 处的一个函数, 其仅仅依赖于纽结而不是辫子本身;

(iv) 确保同一个链接给出同一个函数;

(v) 找到一种方法来计算纽结函数.

纽结和链接

给出纽结的正确定义不是一件容易的事情, 但是可以给出一个直观理解的概念.

定义 1.6.6 一个纽结就是一个从圈 $S_1 \in \mathbb{R}^2$ 到空间 $\mathbb{R}^3$ 上的嵌入 (injective) 映射.

注记 1.6.7 定义中的映射我们要求是连续的, 但这还不足够. 因为有一些讨厌的连续映射 (nasty continuous maps), 它们不能看成是纽结. 如果我们将两条线彼此上下交错的话, 一个纽结是可以画在一张平面上的. 我们把这个性质称为纽结的**投影**(projection).

我们已经知道, 图 6 不是一个纽结, 但其是一个链接 (Borromean 环). 如果移除其中一个环, 它们就全部脱落了.

定义 1.6.8　一个 n-链接就是一个完美的从 n 个互不相交的圈到空间 $\mathbb{R}^3$ 上的嵌入映射 (图 6 和图 7).

图 6　　　　　　　　　　　　　图 7

例 1.6.9　这里给出两个有向链接 (oriented links)(图 8).

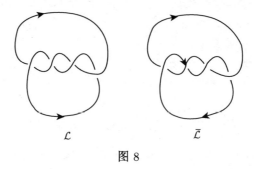

图 8

它们看上去不同, 其实是一样的. 它们都等价于图 9.

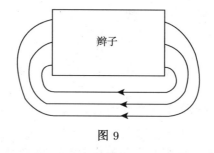

图 9

命题 1.6.10　每个有向链接都可以通过闭合一个辫子得到.

注记 1.6.11　不同的辫子可以得到相同的链接.

定义 1.6.12　设 V 是一个向量空间且 $c \in \mathrm{Aut}(V \otimes V)(V \otimes V$ 到自身的线性

同构集合) 是下面辫子方程

$$(c \otimes \mathbf{id})(\mathbf{id} \otimes c)(c \otimes \mathbf{id}) = (\mathbf{id} \otimes c)(c \otimes \mathbf{id})(\mathbf{id} \otimes c)$$

的解. 称 c 为一个**辫子**, V 为一个**辫子向量空间**.

一个辫子向量空间 V 可以给出一类特殊型的代数: **Nichols 代数**$\mathfrak{B}(V)$.

例 1.6.13 (i) 设 $V = \langle x_1, x_2, \cdots, x_n \rangle$. $c(x_i \otimes x_j) = q_{ij} x_j \otimes x_i$ 对任意 $q_{ij} \in \mathbb{C}^\times$, 那么 c 和后面定义的 Nichols 代数称为是对角型 (diagonal type). 数 $q_{ij}, i, j = 1, 2, \cdots, r$ 定义了一个双特征 (见例 1.1.7):

$$\mathbb{Z}^r \times \mathbb{Z}^r \longrightarrow \mathbb{C}, \quad ((a_1, \cdots, a_r)(b_1, \cdots, b_r)) \mapsto \prod_{i,j=1}^r q_{ij}^{a_i b_j}.$$

(ii) 设 G 是一个群, $V = \mathbb{C}G$. $c(g \otimes h) = ghg^{-1} \otimes g$.

为了定义 Nichols 代数, 我们回顾 Artin 辫子群 $\mathbb{B}_n$. 它是由 $\sigma_1, \sigma_2, \cdots, \sigma_{n-1}$ 所生成的自由群的商群, 商关系为

$$\sigma_i \sigma_{i+1} \sigma_i = \sigma_{i+1} \sigma_i \sigma_{i+1}, \quad 1 \leqslant i \leqslant n-2,$$
$$\sigma_i \sigma_j = \sigma_j \sigma_i, \quad |i-j| \geqslant 2.$$

同时, 回顾群 $\mathbb{S}_n$. 它是由 $\tau_1, \tau_2, \cdots, \tau_{n-1}$ 所生成的自由群的商群, 商关系为

$$\tau_i \tau_{i+1} \tau_i = \tau_{i+1} \tau_i \tau_{i+1}, \quad 1 \leqslant i \leqslant n-2,$$
$$\tau_i \tau_j = \tau_j \tau_i, \quad |i-j| \geqslant 2,$$
$$\tau_i^2 = 1, \quad 1 \leqslant i \leqslant n-1.$$

注记 1.6.14 (i) 存在一个满射 $p: \mathbb{B}_n \longrightarrow \mathbb{S}_n, \sigma_i \mapsto \tau_i$.

(ii) (Matsumoto) 存在集合截面

$$\mu: \mathbb{S}_n \longrightarrow \mathbb{B}_n, \quad \tau_i \mapsto \sigma_i$$

使得 $\mu(xy) = \mu(x)\mu(y)$, 对任意 $x, y \in \mathbb{S}_n$ 满足 $\mathrm{length}(xy) = \mathrm{length}(x) + \mathrm{length}(y)$.

(iii) 设 (V, c) 是一个辫子向量空间, 且

$$c_i = c_{i,i+1} = \mathbf{id}_{V^{\otimes (i-1)}} \otimes c \otimes \mathbf{id}_{V^{\otimes (n-i-1)}} \in \mathrm{Aut}(V^{\otimes n}).$$

那么 $c_1, c_2, \cdots, c_{n-1}$ 满足辫子群关系, 因此

$$\rho_n: \mathbb{B}_n \longrightarrow \mathrm{Aut}(V^{\otimes n}), \quad \sigma_i \mapsto c_i$$

是一个表示.

定义 1.6.15 设 (V,c) 是一个辫子向量空间. 定义 V 的 Nichols 代数为

$$\mathfrak{B}(V) = \bigoplus_n \mathfrak{B}^n(V) = \bigoplus_n T^n(V)/(\mathrm{Ker}\mathfrak{S}_n),$$

这里 $\mathfrak{S}_n$ 是量子对称子 (quantum symmetrizer):

$$\mathfrak{S}_n = \sum_{\sigma \in \mathbb{S}_n} \rho_n(\mu(\sigma)),$$

其中 ρ_n 为 c 诱导出的 $\mathbb{B}_n$ 的表示, μ 是 Matsumoto 截面.

例如: $\mathfrak{S}_2 = 1+c$, $\mathfrak{S}_3 = 1 + c_{12} + c_{23} + c_{12}c_{23} + c_{23}c_{12} + c_{12}c_{23}c_{12}$.

注记 1.6.16 (i) 实际上, 我们定义 $\rho : S_n \longrightarrow \mathrm{End}(V^{\otimes n})$ 如下.
首先, 对一个置换 (transposition) $(i, i+1) \in S_n$, 令

$$\rho((i, i+1)) := \mathbf{id} \otimes \cdots \otimes \mathbf{id} \otimes c \otimes \mathbf{id} \otimes \cdots \otimes \mathbf{id},$$

这里 c 作用在 V 的第 i 和 $i+1$ 项上.

如果 $\omega = \tau_1 \cdots \tau_l$ 是 $\omega \in S_n$ 的一个约化 (reduced) 表示, 那么 $\rho(\omega) := \rho(\tau_1) \cdots \rho(\tau_l)$.

设 $\mathfrak{S}_n := \sum_{\omega \in S_n} \rho(\omega)$. 那么

$$\mathfrak{B}(V) := \bigoplus_{n \geqslant 0} T^n(V)/\mathrm{Ker}(\mathfrak{S}_n)$$

(ii) V. Kharchenko[99] 在 1999 年得到了如下结论.
存在一个全序指标集 $(L, \leqslant)$ 和 $\mathbb{Z}^r$-齐次元素 $X_l \in \mathfrak{B}(V), l \in L$ 使得

$$\{X_{l_1}^{m_1} \cdots X_{l_v}^{m_v} \mid v \geqslant 0, l_1, \cdots, l_v \in L, l_1 > \cdots > l_v, 0 \leqslant m_i < h_{l_v}, \forall i = 1, \cdots, v\}$$

是 $\mathfrak{B}(V)$ 一个向量空间基. 这里

$$h_l = \min\{m \in \mathbb{N} \mid 1 + q_l + \cdots + q_l^{m-1} = 0\} \cup \{\infty\},$$

且 $q_l = \chi(\deg X_l, \deg X_l), l \in L$.

例 1.6.17 (i) 如果 $c(x \otimes y) = y \otimes x, \forall x, y \in V$, 那么 $\mathfrak{B}(V) = S(V)$ 是对称代数 (symmetric algebra);

(ii) 如果 $c(x \otimes y) = -y \otimes x, \forall x, y \in V$, 那么 $\mathfrak{B}(V) = \wedge(V)$ 是外代数 (exterior algebra).

第 2 章　Hopf 代数

本章主要介绍如何用范畴的方法逼近双代数的定义，介绍 Hopf 代数、积分 (integral)、对偶以及相关的作用理论，参考文献 [15, 18, 19, 37, 43, 44, 48, 53, 63, 65, 66, 84, 97, 103, 106, 119, 120, 124, 125, 134—136, 154, 156, 161, 206, 212—222, 235, 250—252, 254, 257, 277, 278].

2.1　双代数的范畴描述

本节主要介绍如何利用双代数 [173] 的模范畴来刻画双代数. 首先设 B 是域 k 上的一个代数，左 B-模范畴 ${}_B\mathcal{M}$ 是 k 上向量空间范畴 Vetc_k 的子范畴. 我们有遗忘函子 $U: {}_B\mathcal{M} \longrightarrow \text{Vetc}_k$.

我们有如下定理.

定理 2.1.1　设 B 是域 k 上的一个代数，那么 B 是双代数当且仅当左 B-模范畴 ${}_B\mathcal{M}$ 是张量范畴且遗忘函子 $U: {}_B\mathcal{M} \longrightarrow \text{Vetc}_k$ 是严格的 (strict) 张量函子, 且 φ_0 和 $\varphi_{M,N}$ 是恒等映射.

证明　遗忘函子 $U: {}_B\mathcal{M} \longrightarrow \text{Vetc}_k$ 是严格的张量函子, 意味着 B-模的张量积就是向量空间的张量积且范畴 ${}_B\mathcal{M}$ 中的单位对象就是 k, 即范畴 Vetc_k 的单位对象.

如果 B 是双代数, 那么畴 ${}_B\mathcal{M}$ 是张量范畴, 存在两个代数同态 $\Delta: B \longrightarrow B \otimes B$ 和 $\varepsilon: B \longrightarrow k$ 满足余结合和余单位性. 设 $M, N \in {}_B\mathcal{M}$, 那么 $M \otimes N$ 的 B-模结构如下:

$$\cdot: B \otimes (M \otimes N) \longrightarrow M \otimes N, \quad a \otimes (m \otimes n) \mapsto a \cdot (m \otimes n) = \sum a_1 m \otimes a_2 n.$$

那么 k 的 B-模结构如下:

$$\cdot: B \otimes k \longrightarrow k, \quad a \otimes \lambda \mapsto a \cdot \lambda = \varepsilon(a)\lambda.$$

很容易验证结合子映射、左单位映射和右单位映射是 B-模同态. 例如, 关于余结合性, 有

$$\sum a_1 \otimes a_{21} \otimes a_{22} = \sum a_1 \otimes (a_2 \cdot (1_B \otimes 1_B))$$
$$= a \cdot (1_B \otimes (1_B \otimes 1_B))$$
$$= a \cdot ((1_B \otimes 1_B) \otimes 1_B)$$
$$= \sum (a_1 \cdot (1_B \otimes 1_B)) \otimes a_2$$
$$= \sum a_{11} \otimes a_{12} \otimes a_2,$$

且对于左单位映射, 有

$$a \cdot l_B(\lambda \otimes m) = a \cdot (\lambda m) = \lambda(a \cdot m)$$
$$= \sum \lambda(\varepsilon(a_1) a_2 \cdot m) = \sum \varepsilon(a_1) \lambda(a_2 \cdot m)$$
$$= \sum l_B((a_1 \cdot \lambda) \otimes (a_2 \cdot m)) = l_B(a \cdot (\lambda \otimes m))$$

对任意 $m \in M, a \in B$.

因此, 畴 $_B\mathcal{M}$ 是张量范畴且遗忘函子 $U: {}_B\mathcal{M} \longrightarrow \text{Vetc}_k$ 是严格的张量函子. 反之, 因为 B 是通过乘法自然称为左 B-模, 所以 $B \otimes B \in {}_B\mathcal{M}$. 定义线性映射

$$\Delta: B \longrightarrow B \otimes B, \quad a \mapsto a \cdot (1_B \otimes 1_B) := \sum a_1 \otimes a_2.$$

很容易验证这个映射是余结合映射. 同时, 有 $\Delta(1_B) = 1_B \cdot (1_B \otimes 1_B) = 1_B \otimes 1_B$.

设 $m \in M, n \in N \in {}_B\mathcal{M}$, 我们可定义 $\rho_m: B \longrightarrow M, a \mapsto a \cdot m$ 和 $\rho_n: B \longrightarrow N, a \mapsto a \cdot n$. 这两个映射显然是 B-模同态. 由张量积的函子性质, 有 $\rho_m \otimes \rho_n: B \otimes B \longrightarrow M \otimes N, a \otimes b \mapsto a \cdot m \otimes b \cdot n$ 也是 B-模同态. 我们有

$$a \cdot (m \otimes n) = a \cdot ((\rho_m \otimes \rho_n)(1_B \otimes 1_B))$$
$$= (\rho_m \otimes \rho_n)(a \cdot (1_B \otimes 1_B))$$
$$= \sum (\rho_m \otimes \rho_n)(a_1 \otimes a_2)$$
$$= \sum (a_1 \cdot m) \otimes (a_2 \cdot n).$$

于是, 可证明 Δ 是代数同态:

$$\Delta(a)\Delta(b) = \sum (a_1 \otimes a_2)(b_1 \otimes b_2) = \sum a_1 b_1 \otimes a_2 b_2$$
$$= \sum \rho_{b_1}(a_1) \rho_{b_2}(a_2) = \sum (\rho_{b_1} \otimes \rho_{b_2})(a_1 \otimes a_2)$$
$$= \sum a \cdot (\rho_{b_1} \otimes \rho_{b_2})(1_B \otimes 1_B) = \sum a \cdot (b_1 \otimes b_2)$$
$$= a \cdot (b \cdot (1_B \otimes 1_B)) = ab \cdot (1_B \otimes 1_B) = \Delta(ab).$$

因为, k 有 B-模结构, 定义线性映射:

$$\varepsilon: B \longrightarrow k, \quad a \mapsto a \cdot 1_k,$$

这定义了余单位. 事实上,

$$(\mathbf{id} \otimes \varepsilon)\Delta(a) = \sum a_1 \otimes \varepsilon(a_2) = \sum r_B(a_1 \otimes \varepsilon(a_2))$$
$$= \sum r_B(a_1 \otimes (a_2 \cdot 1_k)) = \sum r_B \left((\rho_{1_B} \otimes \rho_{1_k}) \sum (a_1 \otimes a_2)\right)$$
$$= r_B((\rho_{1_B} \otimes \rho_{1_k})(a \cdot (1_B \otimes 1_B))) = r_B(a \cdot (1_B \otimes 1_k))$$
$$= a \cdot r_B(1_B \otimes 1_k) = a \cdot 1_B = a1_B = a,$$

类似地, $(\varepsilon \otimes \mathbf{id})\Delta(a) = a$. 最后, 余单位是代数同态:

$$\varepsilon(ab) = ab \cdot 1_k = a \cdot (b \cdot 1_k) = a \cdot (\varepsilon(b)) = \varepsilon(a)\varepsilon(b).$$

对任意 $a, b \in B$. 且显然有: $\varepsilon(1_B) = 1_B \cdot 1_k = 1_k$.

所以, B 有双代数结构.

更一般地, 我们有下面定理.

定理 2.1.2 设 $(\mathcal{C}, \otimes, I, \sigma)$ 是一个 (严格) 辫子张量范畴, 设 B 是该范畴中的一个代数对象, 假设 I 是 $\mathcal{C}$ 中的一个生成子, 且两个函子 $X \otimes \_$ 和 $\_ \otimes X$ 保持满态射和余积 (coproduct) 对任意对象 $X \in \mathcal{C}$. 那么 $(B, \mu_B, \eta_B, \Delta_B, \varepsilon_B)$ 是范畴 $\mathcal{C}$ 中的辫子双代数当且仅当左 B-模范畴 $_B\mathcal{M}$ 具有张量范畴结构且遗忘函子 $U: {_B\mathcal{M}} \longrightarrow \mathcal{C}$ 是严格的张量函子.

注记 2.1.3 (i) 我们将左 B-模范畴换成右 B-模范畴也可以讨论同样问题.

(ii) 我们也可以考虑一个余代数的左或右余模范畴, 得到同样结论.

(iii) 对于辫子张量范畴中, 同样可以考虑双代数结构问题.

例 2.1.4 在域 k 上向量空间范畴 $(\mathrm{Vect}_k, \otimes, k, \tau)$ 中, 下面例子是众所周知的.

(i) 一个群 G 的群代数 kG 是双代数, 具有余乘法 $\Delta(g) = g \otimes g$ 和余单位 $\varepsilon(g) = 1$ 对任意 $g \in G$.

(ii) 设 $R(G)$ 是群 G 的表示函数 (representative function) 集合. 回顾一个函数 $f: G \longrightarrow k$ 称为是可表示的如果存在一个有限维向量空间 V 上的表示 $\pi: G \longrightarrow \mathrm{GL}(V)$, 对 $v \in V, \phi \in V^*$ 使得 $f(g) = \phi(\pi(g)(v))$ 对任意 $g \in G$. 表示函数乘法是点向 (pointwise) 乘法, 因此 $R(G)$ 是交换代数, 且单位元是常函数 $1(g) = 1$, 这是可表示的, 因为有一个平凡表示 $\varepsilon: G \longrightarrow k^\times, \varepsilon(g) = 1$, 此时, $v = \phi = 1 \in k$. 对任意 $f \in R(G)$, 有余乘法 $\Delta(f): G \otimes G \longrightarrow k, (g, h) \mapsto f(gh)$ 和余单位 $\varepsilon: R(G) \longrightarrow k, f \mapsto f(e)$, 这里 e 是群 G 的单位元. 为了证明 $\Delta(f) \in R(G) \otimes R(G)$, 我们考虑有限维表示 $\pi: G \longrightarrow \mathrm{GL}(V)$, 对 $v \in V, \phi \in V^*$ 使得 $f(g) = \phi(\pi(g)(v))$ 对任意 $g \in G$.

取 V 的一组基 $\{e_i\}_{i=1}^n$ 和它的一组对偶基 $\{f^i\}_{i=1}^n$, 那么有

$$\begin{aligned}\Delta(f)(g,h) &= f(gh) = \phi(\pi(gh)(v)) = \phi(\pi(g)(\pi(h)(v))) \\ &= \sum_{i=1}^n \phi(\pi(g)e_i f^i(\pi(h)(v))) = \sum_{i=1}^n \phi(\pi(g)e_i)f^i(\pi(h)(v)) \\ &= \sum_{i=1}^n f_i'(g)f_i''(h),\end{aligned}$$

这里 f_i' 和 f_i'' 是可表示函数, 因为 $f_i'(g) = \phi(\pi(g)e_i)$ 和 $f_i'' = f^i(\pi(h)(v))$. 因此, $\Delta(f) = \sum_{i=1}^n f_i' \otimes f_i'' \in R(G) \otimes R(G)$. 直接证明余乘法和余单位映射是代数同态. 因此, $R(G)$ 是一个双代数.

(iii) 有限群 G 的群代数 kG 的对偶 $(kG)^* = k^G$, 是由元素 $\{p_g\}_{g \in G}$ 生成的, 具有乘法 $p_g p_h = \delta_{g,h} p_g$, 单位 $1 = \sum_{g \in G} p_g$, 余乘法 $\Delta(p_g) = \sum_{hk=g} p_h \otimes p_k$ 和余单位 $\varepsilon(p_g) = \delta_{g,e}$, 于是, k^G 构成了一个双代数.

(iv) 李代数 $\mathfrak{g}$ 的包络代数 $U(\mathfrak{g})$. 其上的余乘法如下给出: 对 $a = X_1 X_2 \cdots X_n$, $X_i \in \mathfrak{g}$,

$$\Delta(a) = a \otimes 1 + 1 \otimes a + \sum_{\sigma \in S_{n,k}} X_{\sigma(1)} \cdots X_{\sigma(k)} \otimes X_{\sigma(k+1)} \cdots X_{\sigma(n)}$$

这里 $S_{n,k} \subset S_n$ 是 (n,k) 个换位 (shuffle) 集合, 即 n 个元素的置换使得 $\sigma(1) < \cdots < \sigma(k)$, 且 $\sigma(k+1) < \cdots < \sigma(n)$. 尤其, 如果 $a = X \in \mathfrak{g}$, 有 $\Delta(X) = X \otimes 1 + 1 \otimes X$. 设 $\Delta(1) = 1 \otimes 1$. 余单位 $\varepsilon(1) = 1$ 和 $\varepsilon(a) = 0$ 对任意 $a = X_1 X_2 \cdots X_n, X_i \in \mathfrak{g}$. 直接证明余乘法和余单位映射是代数同态. 因此, $U(\mathfrak{g})$ 是一个双代数.

(v) 可除幂双代数 $D = \{x_n \mid n \in \mathbb{N}\}$. 其上的代数结构为

$$x_m x_n = \binom{m+n}{m} x_{m+n}, \quad 1_D = x_0,$$

余代数结构为

$$\Delta(x_n) = \sum_{i=0}^n x_i \otimes x_{n-i}, \quad \varepsilon(x_n) = \delta_{n,0}.$$

事实上, 有

$$\Delta(x_m)\Delta(x_n) = \left(\sum_{i=0}^m x_i \otimes x_{m-i}\right)\left(\sum_{j=0}^n x_j \otimes x_{n-j}\right) = \sum_{i,j=0}^m x_i x_j \otimes x_{m-i} x_{n-j}$$

$$= \sum_{i,j=0}^{m} \binom{i+j}{i} x_{i+j} \otimes \binom{(m+n)-(i+j)}{m-i} x_{(m+n)-(i+j)}$$

$$= \sum_{k=0}^{m+n} \sum_{i=0}^{m} \binom{k}{i} \binom{m+n-k}{m-i} x_k \otimes x_{(m+n)-k}$$

$$= \sum_{k=0}^{m+n} \binom{m+n}{m} x_k \otimes x_{(m+n)-k}$$

$$= \Delta\left(\binom{m+n}{m} x_{m+n}\right) = \Delta(x_m x_n).$$

2.2 Hopf 代数定义

我们首先从余环 [33, 35, 36, 79, 83, 112, 225, 287] 开始, 诱导出余代数的定义.

定义 2.2.1 设 C 是一个双 (A,A)-双模.

(i) 一个双 (A,A)-双模同态 $\Delta: C \longrightarrow C \otimes_A C$ 称为是**余乘法**(comultiplication). 对任意 $c \in C$, 记 $\Delta(c) = \sum c_1 \otimes_A c_2$. 余乘 Δ 称为是**余结合**的如果有

$$\sum c_{11} \otimes_A c_{12} \otimes_A c_2 = \sum c_1 \otimes_A c_{21} \otimes_A c_{22}$$

对任意 $c \in C$.

(ii) 一个双 (A,A)-双模同态 $\varepsilon: C \longrightarrow A$ 称为是关于 Δ 的**余单位元**(counit) 如果

$$\sum \varepsilon(c_1)c_2 = c = \sum c_1 \varepsilon(c_2)$$

对任意 $c \in C$.

(iii) 一个 A-余环 C 称为是 A-**余代数**(A-coalgebra) 如果 A 是交换的且左 A-模和右 A-模相同, 即 $ac = ca$ 对任意 $a \in A, c \in C$.

下面我们给出 Hopf 代数的定义.

定义 2.2.2 若五元组 $(H, m, u, \Delta, \varepsilon)$ 满足

(i) $(H, m, 1_H)$ 为域 k 上结合代数.

(ii) (H, Δ, ε) 为域 k 上余代数.

(iii) 对任意 $x, y, z \in H$, 有 $\Delta(xy) = \Delta(x)\Delta(y)$ 和 $\varepsilon(xy) = \varepsilon(x)\varepsilon(y)$(这等价于乘法算子和单位元算子是余代数同态),

则称 H 为**双代数**. 进一步, 我们说 H 是一个**Hopf 代数**如果存在一个元素 $S \in \mathrm{End}_k(H)$ 使得 S 是 id_H 的卷积逆元, 即满足

$$\sum S(h_1)h_2 = \varepsilon(h)1_H = \sum h_1 S(h_2)$$

对任意 $h \in H$, S 称为 H 的**对极**(antipode).

例 2.2.3　(1) **缠绕结构**　设 A 是代数, C 是余代数.

(i) **右-右缠绕结构** (A, C, α)　线性映射 $\alpha : A \otimes C \longrightarrow A \otimes C, a \otimes c \mapsto a_\alpha \otimes c^\alpha$ 满足, 对任意 $a, b, c \in H$

$$(ab)_\alpha \otimes c^\alpha = a_\alpha b_\beta \otimes c^{\alpha\beta}; \tag{2.2.1}$$

$$a_\alpha \varepsilon(c^\alpha) = \varepsilon(c) a; \tag{2.2.2}$$

$$a_\alpha \otimes \Delta(c^\alpha) = a_{\alpha\beta} \otimes c_1{}^\beta \otimes c_2{}^\alpha; \tag{2.2.3}$$

$$1_\alpha \otimes c^\alpha = 1 \otimes c. \tag{2.2.4}$$

设 A 是一个双代数, 线性映射 $\alpha : A \otimes A \longrightarrow A \otimes A$ 定义为 $\alpha(a \otimes b) = (1 \otimes b)\Delta(a)$. 容易验证 (A, A, α) 是一个右-右缠绕结构. 注意, (2.1.1) 用到 Δ 是代数同态和代数的结合性, (2.1.3) 既用到 Δ 是代数同态, 也用到 Δ 的余结合性. 比如: 对于 (2.1.3), 有

$$a_1 \otimes (ca_2)_1 \otimes (ca_2)_2 = a_{11} \otimes c_1 a_{12} \otimes c_2 a_2$$

对任意 $a, c \in H$.

(ii) **左-右缠绕结构** (A, C, α)　线性映射 $\alpha : A \otimes C \longrightarrow A \otimes C, a \otimes c \mapsto a_\alpha \otimes c^\alpha$ 满足, 对任意 $a, b, c \in H$

$$(ab)_\alpha \otimes c^\alpha = a_\alpha b_\beta \otimes c^{\beta\alpha}; \tag{2.2.5}$$

$$a_\alpha \varepsilon(c^\alpha) = \varepsilon(c) a; \tag{2.2.6}$$

$$a_\alpha \otimes \Delta(c^\alpha) = a_{\alpha\beta} \otimes c_1{}^\beta \otimes c_2{}^\alpha; \tag{2.2.7}$$

$$1_\alpha \otimes c^\alpha = 1 \otimes c. \tag{2.2.8}$$

设 A 是一个双代数, 线性映射 $\alpha : A \otimes A \longrightarrow A \otimes A$ 定义为 $\alpha(a \otimes b) = \Delta(a)(1 \otimes b)$. 容易验证 (A, A, α) 是一个左-右缠绕结构.

(iii) **右-左缠绕结构** (A, C, α)　线性映射 $\alpha : C \otimes A \longrightarrow C \otimes A, c \otimes a \mapsto c^\alpha \otimes a_\alpha$ 满足, 对任意 $a, b, c \in H$

$$c^\alpha \otimes (ab)_\alpha = c^{\alpha\beta} \otimes a_\alpha b_\beta; \tag{2.2.9}$$

$$\varepsilon(c^\alpha) a_\alpha = \varepsilon(c) a; \tag{2.2.10}$$

$$\Delta(c^\alpha) \otimes a_\alpha = c_1{}^\alpha \otimes c_2{}^\beta \otimes a_{\alpha\beta}; \tag{2.2.11}$$

$$c^\alpha \otimes 1_\alpha = c \otimes 1. \tag{2.2.12}$$

设 A 是一个双代数, 线性映射 $\alpha : A \otimes A \longrightarrow A \otimes A$ 定义为 $\alpha(a \otimes b) = (a \otimes 1)\Delta(b)$. 容易验证 (A, A, α) 是一个右-左缠绕结构.

(iv) **左-左缠绕结构** (A, C, α) 线性映射 $\alpha : C \otimes A \longrightarrow C \otimes A$, $c \otimes a \mapsto c^\alpha \otimes a_\alpha$ 满足,对任意 $a, b, c \in H$

$$c^\alpha \otimes (ab)_\alpha = c^{\alpha\beta} \otimes a_\beta b_\alpha; \tag{2.2.13}$$

$$a_\alpha \varepsilon(c^\alpha) = \varepsilon(c) a; \tag{2.2.14}$$

$$\Delta(c^\alpha) \otimes a_\alpha = c_1{}^\alpha \otimes c_2{}^\beta \otimes a_{\alpha\beta}; \tag{2.2.15}$$

$$c^\alpha \otimes 1_\alpha = c \otimes 1. \tag{2.2.16}$$

设 A 是一个双代数,线性映射 $\alpha : A \otimes A \longrightarrow A \otimes A$ 定义为 $\alpha(a \otimes b) = \Delta(b)(a \otimes 1)$. 容易验证 (A, A, α) 是一个左-左缠绕结构.

(2) $(B, m, 1_B, \Delta, \varepsilon)$ 设双代数. 考虑 $B \otimes B$ 为双 (B, B)-模如下: 对任意 $a, b, c \in B$

$$(a \otimes b) \cdot c = (a \otimes b)\Delta(c),$$
$$a \cdot (b \otimes c) = ab \otimes c,$$

那么 $A = B \otimes B$ 有 B-余环结构如下:

$$\Delta : B \otimes B \longrightarrow (B \otimes B) \otimes_B (B \otimes B) \simeq (B \otimes B) \cdot 1 \otimes B,$$
$$\Delta(a \otimes b) = \sum (a \otimes b_1) \otimes_B (1 \otimes b_2) = \sum a \otimes b_1 \otimes b_2,$$
$$\varepsilon : B \otimes B \longrightarrow (B \otimes B) \cdot 1 \xrightarrow{\mathrm{id} \otimes \varepsilon} B,$$
$$\varepsilon : a \otimes b \mapsto (a \otimes b) \cdot 1 \mapsto \sum a\varepsilon(b).$$

命题 2.2.4 (1) 设 A 是一个具有单位元的非结合代数且是具有余单位的非余结合余代数. 定义线性映射 $\alpha : A \otimes A \longrightarrow A \otimes A$ 为 $\alpha(a \otimes b) = \Delta(a)(1 \otimes b)$ 和 $\beta : A \otimes A \longrightarrow A \otimes A$ 定义为 $\beta(a \otimes b) = (a \otimes 1)\Delta(b)$. 那么 A 是一个双代数当且仅当 (A, A, α) 是一个左-右缠绕结构, 且 (A, A, β) 是一个右-左缠绕结构. 更进一步, A 是一个 Hopf 代数当且仅当 $\alpha : A \otimes A \longrightarrow A \otimes A$, $\beta : A \otimes A \longrightarrow A \otimes A$ 是双射.

(2) $(B, m, 1_B, \Delta, \varepsilon)$ 和 $(B, m, 1_B, \Delta^{\mathrm{cop}}, \varepsilon^{\mathrm{cop}})$ 诱导 $B \otimes B$ 上的 B-余环当且仅当 B 是双代数. 进一步, 映射 $\beta : B \otimes B \longrightarrow B \otimes B$ 定义为 $\beta(a \otimes b) = (a \otimes 1)\Delta(b)$ 是双射当且仅当 B 是 Hopf 代数.

证明 (1) 由例 2.1.3(ii), (iii), 充分性显然. 关于必要性, 由 (2.1.5), 容易看出 Δ 是代数同态且乘法是结合的; 由 (2.1.7), 可得 A 是余结合余代数.

(2) 易证.

我们可以自然地给出同态和理想的概念. 映射 $f : H \to K$ 是一个 **Hopf** 同态(Hopf morphism) 如果它是一个双代数同态, 并且 $f(S_H h) = S_K f(h)$. H 的一个

子空间 I 是一个**Hopf 理想**(Hopf ideal) 如果它是一个双理想并且 $S(I) \subseteq I$; 在这种情况下 H/I 可以从 H 上诱导出**Hopf 代数**结构.

例 2.2.5 (i) 设 $H = KG$ 是一个群代数, 如果定义 $Sg = g^{-1}, g \in G$, 那么 H 是一个 Hopf 代数. 进一步地, 对任意的 Hopf 代数 H 和 $g \in G(H)$, 可以推出 $Sg = g^{-1}$; 特别地, 所有的群像元素在 H 中都是可逆的.

设 p 是一个素数, 那么维数为 p 的 Hopf 代数同构于群代数 KG_p, 这里 G_p 表示域 K 中 1 的 p 次根群 [284].

(ii) 设 $H = U(g)$ 是一个**包络代数**. 如果定义 $Sx = -x, x \in g$, 那么 H 是一个 Hopf 代数. 进一步地, 对任意的 Hopf 代数 H 和 $x \in P(H)$, 可以推出 $Sx = -x$.

读者可以验证一下, 如果 $x \in P_{g,h}(H)$, 我们一定有 $Sx = -h^{-1}xg^{-1}$.

(iii) 如果 H 是一个 Hopf 代数, 那么 Sweedler 对偶 H^o 也是一个 Hopf 代数, 并且它的对极是 S^*.

例 2.2.6 Sweedler 四维 Hopf 代数是最小的非交换的、非余交换的 Hopf 代数, 并且在给定特征不等于 2 的域 K 的意义下是唯一的 (也是**自对偶**的). 我们描述它如下:

$$H_4 = K\langle 1, g, x, gx | g^2 = 1, x^2 = 0, xg = -gx\rangle,$$
$$\Delta g = g \otimes g, \quad \Delta x = x \otimes 1 + g \otimes x,$$
$$\varepsilon(g) = 1, \quad \varepsilon(x) = 0,$$
$$Sg = g = g^{-1}, \quad Sx = -gx \quad (g \in G(H), \ x \in P_{1,g}(H), \dim(S) = 4).$$

例 2.2.7 (i) 作为 Sweedler 四维 Hopf 代数推广, Taft 已经构造出了无穷多个对极的秩为 $2n\ (n > 1)$ 的有限维 Hopf 代数 [12]. 一般地, 设 $q \in K$ 是阶数 $N \geqslant 2$ 的 1 的一个根. Taft 代数 $T_K(q) = T(q)$ 是一个代数 $K\langle g, x \mid gxg^{-1} = qx, g^N = 1, x^N = 0\rangle$. 它的 Hopf 代数结构为

$$\Delta(g) = g \otimes g, \quad S(g) = g^{-1}, \quad \varepsilon(g) = 1,$$
$$\Delta(x) = x \otimes g + 1 \otimes x, \quad S(x) = -xg^{-1}, \quad \varepsilon(x) = 0.$$

$T(q)$ 的维数是 N^2. 那么, 作为 Hopf 代数, 有 $T(q) \simeq T(q)^*$, 而且 $T(q) \simeq T(q')$ 当 $q = q'$.

尤其, 阶数 (维数) 为 p^2 的半单 Hopf 代数是群代数 [130]; 只有维数为 p^2 的有点非余半单 Hopf 代数为 Taft 代数 $T_K(q) = T(q), q \in G_p - 1$.

(ii) 设 $q \in G_p - 1$, 那么域 K 上维数为 p^3 的不同构的全部有点非余半单 Hopf 代数如下 [12]:

(a) 张量积 Hopf 代数 $T(q) \otimes K\mathbb{Z}/(p)$.

2.2 Hopf 代数定义

(b) Hopf 代数 $\widetilde{T(q)} := K\langle g, x \mid gxg^{-1} = q^{\frac{1}{p}}x, g^{p^2} = 1, x^p = 0\rangle$. 因此 $q^{\frac{1}{p}}$ 是 q 的 p 次根. 它的余积为: $\Delta(x) = x \otimes g^p + 1 \otimes x, \Delta(g) = g \otimes g$.

(c) Hopf 代数 $\widehat{T(q)} := K\langle g, x \mid gxg^{-1} = qx, g^{p^2} = 1, x^p = 0\rangle$. 它的余积为: $\Delta(x) = x \otimes g + 1 \otimes x, \Delta(g) = g \otimes g$.

(d) Hopf 代数 $R(q) := K\langle y, x \mid yxy^{-1} - q^{\frac{1}{p}}x, y^{p^2} - 1, x^p - 1 - g^p\rangle$. 它的余积为: $\Delta(x) = x \otimes g + 1 \otimes x, \Delta(g) = g \otimes g$.

(e) Frobenius-Lusztig 核 $u(q) := K\langle g, x, y \mid gxg^{-1} = q^2x, gyg^{-1} = q^{-2}y, g^p = 1, x^p = y^p = 0, xy - yx = g - g^{-1}\rangle$. 它的余积为: $\Delta(x) = x \otimes g + 1 \otimes x, \Delta(y) = y \otimes 1 + g^{-1} \otimes y, \Delta(g) = g \otimes g$.

(f) 设 $m \in \mathbb{Z}/(p) - 0$, 那么书 (book)Hopf 代数 $b(q,m) := K\langle g, x, y \mid gxg^{-1} = qx, gyg^{-1} = q^m y, g^p = 1, x^p = y^p = 0, xy - yx = 0\rangle$. 它的余积为: $\Delta(x) = x \otimes g + 1 \otimes x, \Delta(y) = y \otimes 1 + g^m \otimes y, \Delta(g) = g \otimes g$.

例 2.2.8 双代数 $B = \mathcal{O}(M_n(K)) = K[X_{ij} | 1 \leqslant i, j \leqslant n]$($n \times n$ 矩阵上的多项式函数) 不是一个 Hopf 代数, 因为群像元素 $\det X$ 在 B 中不可逆. 但是, 有一些 Hopf 代数与 B 密切相关:

(a) $\mathcal{O}(\mathrm{SL}_n(K)) = \mathcal{O}(M_n(K))/(\det X - 1)$;

(b) $\mathcal{O}(\mathrm{GL}_n(K)) = \mathcal{O}(M_n(K))[(\det X)^{-1}]$.

如果定义 $SX = X^{-1}$, 这些双代数就变成 Hopf 代数了; 也就是说 SX_{ij} 是 X^{-1} 的 ij 次方的对象.

例如, 设 $M(2) = k[X_{11}, X_{12}, X_{21}, X_{22}]$ 是一个由 X_{11}, X_{12}, X_{21} 和 X_{22} 生成的一个交换代数, 那么 $M(2)$ 是一个双代数具有余结构为

$$\Delta(X_{ij}) = \sum_{l=1}^{2} X_{il} \otimes X_{lj}, \quad \varepsilon(X_{ij}) = \delta_{ij}.$$

如果 X 是一个非退化矩阵 (generic matrix)(X_{ij}), 那么行列式 (determinant)$d = \det(X)$ 是一个群像元素, 即 $d \neq 0, \Delta(d) = d \otimes d$. 设 $\mathrm{GL}(2) = M(2)[y]/(dy - 1)$. 那么 $\mathrm{GL}(2)$ 是一个 Hopf 代数具有反对极: $S(X) = X^{-1}$, 即

$$S(X_{11}) = yX_{22}, \quad S(X_{12}) = -yX_{12},$$
$$S(X_{21}) = -yX_{21}, \quad S(X_{22}) = yX_{11}, \quad S(y) = d.$$

对 $0 \neq q \in k$, 双代数 $M_q(2)$ 是变量 $X_{ij}, i, j = 1, 2$ 上的一个非交换自由代数, 且在变量上具有如上的余乘法和余单位. 但是变量满足如下六个关系式:

$$X_{12}X_{11} = qX_{11}X_{12}, \quad X_{21}X_{11} = qX_{11}X_{21}, \quad X_{22}X_{12} = qX_{12}X_{22},$$
$$X_{22}X_{21} = qX_{21}X_{22}, \quad X_{21}X_{12} = X_{12}X_{21}, \quad X_{22}X_{11} = X_{11}X_{22} + (q - q^{-1})X_{12}X_{21}.$$

量子行列式 $d = X_{11}X_{22} - q^{-1}X_{21}X_{12} = X_{22}X_{11} - qX_{12}X_{21}$ 是群像元素且是中心元. 类似地, $\mathrm{GL}_q(2) = M_q(2)[y]/J$, J 是由 $dy - 1$ 所生成的 $M_q(2)[y]$ 的理想, 是一个 Hopf 代数, 具有反对极为

$$S(X) = X^{-1} = y \begin{pmatrix} X_{22} & -qX_{12} \\ -q^{-1}X_{21} & X_{11} \end{pmatrix},$$

且 $S(y) = d$.

定义 2.2.9 设 $\mathbb{K}$ 是一个特征为 0 的代数闭域, $\mathbb{K}^\times = \mathbb{K} - \{0\}$, G 是一个有限 Abelian 群. 一个 G 上的**双特征**(bicharacter) 就是一个 $\mathbb{Z}$ 双线性映射 (bilinear form)$\beta : G \times G \to \mathbb{K}^\times$, 即对任意 $g, h, l \in G$

$$\beta(gh, l) = \beta(g, l)\beta(h, l), \quad \beta(g, hl) = \beta(g, h)\beta(g, l).$$

注记 2.2.10 (i) 设 β 是 G 上的双特征, 则 $\mathrm{Vect}^G = \{V = \oplus_{g \in G} V_g\}$ 是一个辫子范畴, 辫子定义如下, 对任意 $v \in V_g, w \in W_h$,

$$c = c_\beta : V \otimes W \to W \otimes V, \quad c(v \otimes w) = \beta(g, h)w \otimes v.$$

(ii) 我们用 Vect^G_β 表示辫子为 c_β 的辫子张量范畴 Vect^G.

定义 2.2.11 设 G 是一个有限 Ablian 群, 如果 B 是范畴 Vect^G_β 中的双代数, 即 $B = \bigoplus_{g \in G} B_g$ 是一个 G-分次代数和余代数使得 $\Delta : B \to B \underline{\otimes} B$ 和 ε 都是 G-分次代数同态, 那么我们称 B 是一个**有色双代数**(color bialgebra). 如果 Id_B 有卷积逆 $\mathcal{S} \in \mathrm{End}(H)$ 使得

(i) $\mathcal{S}(xy) = \beta(|x|, |y|)\mathcal{S}(y)\mathcal{S}(x)$;

(ii)$\Delta(\mathcal{S}(x)) = \beta(|x^{(1)}|, |x^{(2)}|)\mathcal{S}(x^{(2)}) \otimes \mathcal{S}(x^{(2)})$.

对任意 $x, y \in B$ 都成立, 那么我们称 B 是一个**有色 Hopf 代数**(color Hopf algebra).

注记 2.2.12 粗略地讲, 一个有色 Hopf 代数就是一个 G-分次代数和一个 G-分次余代数并且满足一些兼容性条件, 因此我们也可以定义有色 Hopf 代数如下.

定义 2.2.13 一个**有色 Hopf 代数**A 是一个 6-元组 $(A, m, u, \Delta, \varepsilon)$ 使得

(i) $A = \bigoplus_{g \in G} A_g$ 是一个 G-分次代数、分次余代数.

(ii) Δ, ε 都是代数同态, 即对任意 $a, b \in A$, 有

$$\varepsilon(ab) = \varepsilon(a)\varepsilon(b),$$
$$\Delta(ab) = \Sigma\chi(|a_2|, |b_1|)a_1b_1 \otimes a_2b_2,$$

其中, $\chi : G \times G \to \mathbb{K}^\times$ 是 G 上的一个双特征.

2.2 Hopf 代数定义

(iii) 对极 S 是一个分次映射使得

$$\Sigma a_1 S a_2 = \varepsilon(a) = \Sigma S a_1 a_2$$

对任意的齐次元 $a \in A$.

命题 2.2.14 对极保持次数, 即 $|Sa| = |a|$, 对任意的齐次元 (homogenous) $a \in A$.

命题 2.2.15 设 A 是一个有色 Hopf 代数, 则对任意 $a, b \in A$, 对极 S 满足

$$S(ab) = \chi(|a|, |b|) S(b) S(a),$$
$$\Delta(Sa) = \Sigma \chi(|a_1|, |a_2|) S a_2 \otimes a_1.$$

定义 2.2.16 设 A 是一个有色 Hopf 代数, 一个**分次右 A-有色 Hopf 模**(graded right A-color Hopf module) 是一个分次 $\mathbb{K}$ 空间 M 使得

(i) M 是一个分次右 A-模.
(ii) M 是一个分次右 A-余模 ($\rho(m) = \Sigma m_0 \otimes m_1$).
(iii) ρ 是一个右 A-模同态, 即

$$\rho(ma) = \Sigma \chi(|m_1|, |a_1|) m_0 a_1 \otimes m_1 a_2.$$

例 2.2.17 设 M 是一个分次 $\mathbb{K}$ 空间. 我们在 $M \otimes A$ 上定义一个分次右 A-模, 对任意 $m \in M, a, b \in A$, $(m \otimes a)b = m \otimes ab$. 在 $M \otimes A$ 上定义一个分次右 A-余模, $\rho(m \otimes a) = \Sigma m \otimes a_1 \otimes a_2$. 我们容易验证 $M \otimes A$ 是一个分次右 A-有色 Hopf 模.

命题 2.2.18 设 A 是一个有色 Hopf 代数, 如果 $a, b \in A$ 都是齐次元, 那么

$$\varepsilon(a) \chi(|a|, |b|) = \varepsilon(a).$$

命题 2.2.19 设 A 是一个有色 Hopf 代数, M 是一个右 A-有色 Hopf 模. 则 $M \cong M^{\text{co}A} \otimes A$ 是一个右有色 Hopf 模, 其中 $M^{\text{co}A} \otimes A$ 是一个平凡的右有色 Hopf 模. 特别地, M 是一个自由右 A-有色 Hopf 模.

命题 2.2.20 如果 $\beta = \varepsilon$ 是一个平凡的双特征, 那么一个 (G, ε)-有色 Hopf 代数是一个 G-分次 Hopf 代数. 一个有限维有色 Hopf 代数 H 的对偶仍然是一个有色 Hopf 代数.

命题 2.2.21 设 H 是一个 Hopf 代数, 那么 H 是一个 G-分次 Hopf 代数当且仅当存在一个同态 $\varphi : A \to \text{Aut} H$, 显然, $\text{Stab}_A(H) = \langle \text{Sub} H \rangle^\perp$. 因此, H 是一个 (G, β)-有色 Hopf 代数当且仅当 $\beta|_{\sup H \times \sup H} = 1$, 当且仅当 $\chi(\sup H) < \text{Stab}_A(H)$.

命题 2.2.22 设 Γ 是一个有限群, 由命题 2.2.21 有 $K\Gamma$ 是一个 G-分次 Hopf 代数当且仅当存在一个 A 群自同态作用在 Γ 上. 固定这个作用, G 也通过 χ 作用

在 Γ 上. 显然, $\operatorname{Stab}_A(K\Gamma) = \operatorname{Stab}_A(\Gamma)$, 因此 β 在 sub$K\Gamma$ 上是平凡的当且仅当

$$\operatorname{sub}K\Gamma <_{\chi^{-1}} (\operatorname{Stab}_A(\Gamma)) = \operatorname{Stab}_G(\Gamma) = \cap_{\eta \in \Gamma} G^{\eta}.$$

例 2.2.23 李超代数的泛包络代数、有色李代数的泛包络代数都是有色 Hopf 代数.

例 2.2.24 设 $G = C_4 = \langle g \rangle$ 是一个秩为 4 的循环群, $i \in K$ 使得 $i^2 = -1$ 和 $\beta : G \times G \to K^{\text{times}}, \beta(g^i, g^j) = (-1)^{ij}$, 那么 $A = \langle a \rangle$, 其中 $a(g) = i$. 设 $\Gamma = C_2 \oplus C_2 = \langle \gamma \rangle \oplus \langle \eta \rangle$, 定义 A 的作用为 $a(\gamma) = \gamma, a(\eta) = \gamma + \eta$. 那么 $K\Gamma$ 是一个有色 Hopf 代数, 因为对任意 $x, y \in \sup K\Gamma = \{1, g^2\}$, 有 $\beta(x,y) = 1$.

例 2.2.25 设 $G = C_2 \oplus C_4 = \langle g \rangle \oplus \langle h \rangle$, $i \in K$ 使得 $i^2 = -1$ 和 $\beta : G \times G \to K^{\times}, \beta(g^i + h^i, g^k + h^l) = i^{il-jk}$, 那么 $A = \langle a \rangle \oplus + \langle b \rangle$, 其中 $a(g) = -1, a(h) = b(g) = 1, b(h) = 1$. 设 $\Gamma = C_2 \oplus C_2 = \langle \gamma \rangle \oplus \langle \eta \rangle$, 定义 A 的作用为 $a(\gamma) = b(\gamma) = \gamma, a(\eta) = b(\eta) = \gamma + \eta$. 那么 $K\Gamma$ 是一个有色 Hopf 代数, 因为对任意 $x, y \in \sup K\Gamma = \{1, g + h^2\}$, 有 $\beta(x, y) = 1$.

定义 2.2.26 设 C 是一个余代数, C_0 表示它的**余根**(coradical), 即 C 的所有单子余代数的和. C 称为是**连通**(connected) 如果 C_0 是一维的. 如果每个单子余代数是一维的, 则 C 称为是**有点的**(pointed). 一个双代数 B 是**有点的**(**连通的**), 如果它作为余代数是有点的 (连通的). 如果 B 是一个有点双代数, 那么 $G(B)$ 在 B 的乘法下是一个幺半群 (monoid). 一个有点双代数 B 是一个**有点 Hopf 代数**(pointed Hopf algebra) 当且仅当 $G(B)$ 在 B 的乘法下是一个群. 一个连通双代数 B 是一个**连通 Hopf 代数**(connected Hopf algebra), 如果 B_0 是一维的.

命题 2.2.27 设 H 是有点的 Hopf 代数. 那么

(i) H 的每一个子余代数都是一维的, 并且每个子群代数都具有形式 $Kg, g \in G(H)$;

(ii) H_0 是由群像元素生成的群代数 KG, 并且是 H 的子 Hopf 代数;

(iii) 我们有

$$H_1 = H_0 \oplus (\oplus_{g,h \in G} \mathcal{P}'_{g,h}),$$

其中 $\mathcal{P}_{g,h}(H) = \{x \in H | \Delta(x) = x \otimes g + h \otimes x\}$, 其中 $\mathcal{P}'_{g,h}$ 是 $\mathcal{P}_{g,h}$ 的子空间, 并且 $\mathcal{P}_{g,h} = \mathcal{P}'_{g,h} \oplus K(G - H)$.

注记 2.2.28 (i) 如果 C 是有点余代数, 则 $C_0 = KG(C)$.

(ii) 设 H 是一个余根滤 (coradical filtration) 为 $\{H_n\}_{n \geq 0}$ 点 Hopf 代数, 则结合的分次空间 $\operatorname{gr}H = \oplus_{n \geq 0} H_n/H_{n-1}$ 是一个分次 Hopf 代数.

(iii) 所有的有限维有点 Hopf 代数的分类问题首先都可以看成是有限维 Hopf 代数的分类, 最近的结果可见文献 [13, 69].

例 2.2.29 (i) (S_n 上的有点 Hopf 代数) 设 $\mathcal{H}(\mathcal{Z}_n^{-1}[t])$ 是由 $\{a_i, H_r : i \in \mathcal{O}_2^n, r \in S_n\}$ 通过下面关系生成的代数:

$$\begin{cases} H_e = 1, \quad H_r H_s = H_{rs}, \; r, s \in S_n, \\ H_j a_i = -a_{jij} H_j, \; i, j \in \mathcal{O}_2^n, \\ a_{(12)}^2 = 0, \\ a_{(12)} a_{(34)} + a_{(34)} a_{(12)} = \alpha(1 - H_{(12)} H_{(34)}), \\ a_{(12)} a_{(23)} + a_{(23)} a_{(13)} + a_{(13)} a_{(12)} = \beta(1 - H_{(12)} H_{(23)}). \end{cases}$$

(ii) 设 $\mathcal{H}(\mathcal{Z}_n^\chi[\lambda])$ 是由 $\{a_i, H_r : i \in \mathcal{O}_2^n, r \in S_n\}$ 通过下面关系生成的代数:

$$\begin{cases} H_e = 1, \quad H_r H_s = H_{rs}, \; r, s \in S_n, \\ H_j a_i = \chi_i(j) a_{jij} H_j, \; i, j \in \mathcal{O}_2^n, \\ a_{(12)}^2 = 0, \\ a_{(12)} a_{(34)} - a_{(34)} a_{(12)} = 0, \\ a_{(12)} a_{(23)} + a_{(23)} a_{(13)} + a_{(13)} a_{(12)} = \beta(1 - H_{(12)} H_{(23)}). \end{cases}$$

(iii) 设 $\mathcal{H}(\mathcal{D}[t])$ 是由 $\{a_i, H_r : i \in \mathcal{O}_4^4, r \in S_4\}$ 生成关系式如下:

$$\begin{cases} H_e = 1, \quad H_r H_s = H_{rs}, \; r, s \in S_n, \\ H_j a_i = -a_{jij} H_j, \quad i \in \mathcal{O}_4^4, \quad j \in \mathcal{O}_2^4, \\ a_{(1234)}^2 = \alpha(1 - H_{(13)} H_{(24)}), \\ a_{(1234)} a_{(1432)} + a_{(1432)} a_{(1234)} = 0, \\ a_{(1234)} a_{(1243)} + a_{(1243)} a_{(1423)} + a_{(1423)} a_{(1234)} = \beta(1 - H_{(12)} H_{(13)}). \end{cases}$$

定义 2.2.30 设 $(H, m, u, \Delta, \varepsilon, S)$ 是一个 Hopf 代数, 如果

(i) $H = \bigoplus_{i=0}^\infty H(i)$ 是一个分次代数[29].

(ii) $H = \bigoplus_{i=0}^\infty H(i)$ 是一个分次余代数.

(iii) $S(H(n)) \subset H(n)$, 对任意 $n \geqslant 0$,

那么我们称 H 是一个**分次 Hopf 代数**(graded Hopf algebra). 进一步地, 如果 $H = \bigoplus_{i=0}^\infty H(i)$ 是一个**余根分次余代数**(coradically graded coalgebra), 那么 H 是一个**余根分次 Hopf 代数**(coradically graded Hopf algebra).

注记 2.2.31 余根分次 Hopf 代数的概念是自然的. 例如, 设 H 是一个余根滤为 $\{H_n\}_{n \geqslant 0}$ 有点 Hopf 代数, 则结合的分次空间 $\mathrm{gr} H = \oplus_{n \geqslant 0} H_n / H_{n-1}$ 是一个分次 Hopf 代数. 进一步地, $\mathrm{gr} H$ 是一个余根分次余代数. 因此, $\mathrm{gr} H$ 是一个余根分次 Hopf 代数.

例 2.2.32 设 B 是一个 Hopf 代数, M 是一个 B-Hopf 双模. 那么 M 在 B 中的张量代数 $T_B(M)$ 是一个分次代数, M 在 B 中的张量余代数 $T_B^B(M)$ 是一个分次余代数, 并且 $T_B(M)$ 和 $T_B^B(M)$ 都是分次 Hopf 代数.

例 2.2.33　一个 $(G, \beta = \varepsilon)$-有色 Hopf 代数就是一个 G-分次 Hopf 代数.

引理 2.2.34　设 C 是一个余代数, 则存在一个具有如下性质的 Hopf 代数 $\mathcal{H}(C)$:

(i) 存在余代数映射 $i : C \to \mathcal{H}(C)$.

(ii) 对任意的 Hopf 代数 H 和余代数映射 $f : C \to H$, 存在一个 Hopf 代数映射 $f' : \mathcal{H}(C) \to H$ 使得 $f = f'i$.

定义 2.2.35　Hopf 代数 $\mathcal{H}(C)$ 称为由 C 生成的**自由 Hopf 代数**(free Hopf algebra).

例 2.2.36　设 $V = \bigoplus_{i=0}^{\infty} V_i$, 其中, 如果 i 是偶数, 那么 $V_i = V$; 如果 i 是奇数, 那么 $V_i = C^{\mathrm{cop}}$. 我们注意到 V 有一个自然的余代数结构.

设 $S : V \to V^{\mathrm{cop}}$ 是一个余代数映射, 把 $(x_0, x_1, x_2, \cdots)$ 映射到 $(0, x_0, x_1, x_2, \cdots)$, 那么 S 诱导出一个双代数映射 $S : T(V) \to T(V)^{\mathrm{op,cop}}$. 设 I 是由集合

$$\{S * \mathrm{Id}(x) - \varepsilon(x)1 | x \in V\} \cup \{\mathrm{Id} * S(x) - \varepsilon(x)1 | x \in V\}$$

生成的 $T(V)$ 的双边理想. 其中, $*$ 代表卷积. 而且, I 是余理想, $S(I) \subset I$. Hopf 代数 $T(v)/I$ 加上从映射 S 诱导出的对极, 就是一个泛对象 $\mathcal{H}(C)$.

命题 2.2.37　设 C 是一个有点余代数, 对任意 $n \geqslant 1$, 下面的表述都是真的:

(i) Hopf 代数 $\mathcal{H}(C)$ 是有点的并且 $\mathcal{H}(C)$ 的余有理化 (coradical) 与由 $G(C)$ 以及它们的逆生成的子代数是相等的, 同时与 $K\langle G(C)\rangle$ 同构.

(ii) 对任意 $n \geqslant 1$, C^n 是 $\mathcal{H}(C)$ 的子余代数, $G(C^n)$ 中每个元素都可以表示成至少 n 个 $G(C)$ 中元素的乘积.

命题 2.2.38　设 $\langle G(C)\rangle$ 是由集合 $G(C)$ 生成的自由群, 则 $\mathcal{H}(kG(c)) = k\langle G(C)\rangle$.

定义 2.2.39　如果 B 是一个 Hopf 代数, M 是一个 B-Hopf 双模, 那么我们说 (B, M) 是一个 Hopf 双模. 进一步地, 如果 M 是一个 Yetter-Drinfel'd 范畴 ${}_B^B\mathcal{YD}$ 中的辫子 Hopf 代数, 我们说 (B, M) 是一个**Yetter-Drinfel'd Hopf 代数**.

定义 2.2.40　设 A 是带有具有齐次次数为 1 的微分 d 的一个 Hopf 代数, 使得下图

$$\begin{array}{ccccc} A & \xrightarrow{\psi} & A \otimes A & \xrightarrow{\varphi} & A \\ \downarrow d & & \downarrow \bar{d} & & \downarrow d \\ A & \xrightarrow{\psi} & A \otimes A & \xrightarrow{\varphi} & A \end{array}$$

是交换的. 那么 A 称为**微分 Hopf 代数**(differential Hopf algebra). A 诱导出的**上同调代数**(cohomology algebra)$H^*(A)$ 也是一个 Hopf 代数.

命题 2.2.41　设 A 是一个带有微分 d 且具有微分次数为 1 的微分 Hopf 代数, 则 A^* 也是微分 Hopf 代数, 它的次数为 -1 的微分子 ∂ 定义为: 对任意

$x \in A, a \in A^*$,
$$\langle x, \partial a \rangle = \langle dx, a \rangle$$

对任意 $u \in A^*$, 定义一个线性映射
$$\theta_u : A \to A$$

具有次数 $-i$ 为
$$\theta_u x = \Sigma_t \langle x_t, u \rangle y_t,$$

其中 $\psi(x) = \Sigma_t x_t \otimes y_t$.

命题 2.2.42 对任意 $c \in A^*$, 设 $[\theta_c, d] = \theta_c \circ d - (-1)^{degc} d \circ \theta_c$, 则 $[\theta_c, d] = \theta_{\partial c}$.

引理 2.2.43 设 $\Im \subset U_q(\mathfrak{g})$ 使得 $\Im$ 生成 $U_q(\mathfrak{g}. \Lambda(\Im) \rtimes D(U_q(\mathfrak{g})))$ 是一个分次超 -Hopf 代数.

引理 2.2.44 设 $\phi := (\mathbf{id}_\Lambda \otimes \Delta_U) : \Lambda(\Im) \otimes U_q(\mathfrak{g}) \to \Lambda(\Im) \otimes D(U_q(\mathfrak{g}))$, 则 $\phi(\Lambda(\Im) \otimes U_q(\mathfrak{g}))$ 是 $\Lambda(\Im) \otimes D(U_q(\mathfrak{g}))$ 的一个子 Hopf 代数, 因此 $\Lambda(\Im) \otimes U_q(\mathfrak{g})$ 通过嵌入映射 ϕ 变成一个超-Hopf 代数. 用 $m_K, \eta_K, \Delta_K, \varepsilon_K$ 来表示 $\Lambda(\Im) \otimes U_q(\mathfrak{g})$ 的 Hopf 代数结构映射. 代数 $\Lambda(\Im) \otimes U_q(\mathfrak{g}), m_K, \eta_K$ 和代数 $K(q, \mathfrak{g})$ 是一致的, 因此空间 $(K(q, \mathfrak{g})), \Delta_K, \varepsilon_K, S_K$ 是一个超 -Hopf 代数.

定理 2.2.45 复形 $(K(q, \mathfrak{g}), d)$ 结合引理 2.1.27 中定义的分次 Hopf 代数结构是一个微分 Hopf 代数.

注记 2.2.46 关于 Hopf*-代数概念见文献 [185].

2.3 积 分

Hopf 代数的积分 [131, 132] 概念来自于高等数学中微积分概念, 同时又比其更加抽象, 决定着 Hopf 代数的一些重要结构 [20, 28, 38, 61, 150, 166, 253, 264, 283].

这里收集一些信息, 关于 Hopf 代数中的积分概念由 Sweedler 引进 ([173, 174]). 进一步, 一个主理想整环上的自由有限维双代数是一个 Hopf 代数当且仅当它拥有一个非退化的左积分 (Larson-Sweedler 定理); 一个主理想整环上的自由有限维 Hopf 代数的反对极是双射; 域上一个 Hopf 代数是有限维的当且仅当它拥有一个非零左积分; 域上有限维 Hopf 代数中左积分形成一个一维子空间; 域上一个 Hopf 代数是半单的当且仅当它拥有一个正则左积分 (Maschke 定理).

定义 2.3.1 (i) 设 B 为一个 Hopf 代数, 我们称 $(M, \cdot, \rho)$ 为 B 上的一个右右B-**Hopf 模**, 若 $(M, \cdot)$ 为一个右 B-模, 而 (M, ρ^r) 为右 B-余模, 并且满足下面相容条件:
$$\rho^r(m \cdot a) = m_0 \cdot a_1 \otimes m_{(1)} a_2$$

对于任意 $a \in B, m \in M$. 其中 $\rho^r(m) = m_0 \otimes m_{(1)}$.

(ii) 一个四角 B-Hopf 双模 (M, ρ^l, ρ^r) 既是一个 B-双模同时也是一个 B-双余模, 使得 ρ^l, ρ^r 是 B-双模同态.

注记 2.3.2 (i) 对一个右 B-余模 (M, ρ^r), 记 B-不变集合为 $M_r^{coB} = \{x \mid \rho^r(x) = x \otimes 1_B\}$. 类似地, 对一个左 B-余模 (M, ρ^l) 有: $M_l^{coB} = \{x \mid \rho^l(x) = 1_B \otimes x\}$.

(ii) 由 B 上的右右 B-Hopf 模对象以及模同态和余模同态所组成的范畴记为 $\mathcal{H}_B^B$.

(iii) 类似可以定义左右 B-Hopf 模范畴 ${}_B\mathcal{H}^B$、左左**B-Hopf 模**${}_B^B\mathcal{H}$、右左B-**Hopf 模** ${}^B\mathcal{H}_B$ 和四角 B-Hopf 双模范畴 ${}_B^B\mathcal{H}_B^B$.

例 2.3.3 对一个四角 B-Hopf 双模 (M, ρ^l, ρ^r), 容易检查 M_l^{coB} 是一个子右 B-余模, 且有一个右 B-模结构

$$m \cdot b = S(b_1) m b_2.$$

类似地, M_r^{coB} 是一个子左 B-余模, 且有一个左 B-模结构

$$b \cdot m = b_1 m S(b_2)$$

对任意 $m \in M, b \in B$. 进一步, M_l^{coB} 和 M_r^{coB} 是 Yetter-Drinfel'd 模范畴.

命题 2.3.4 (Hopf 模基本定理) 设 B 是一个 Hopf 代数. $M \in \mathcal{H}_B^B$, 那么有下面的 B-Hopf 模同构:

$$\alpha : M^{coB} \otimes B \cong M, \quad x \otimes b \mapsto x \cdot b.$$

带有逆映射: $\alpha^{-1}(m) = m_0 \cdot S(m_{(1)}) \otimes m_{(2)}$, 对任意 $m \in M, x \in M^{coB}, b \in B$.

定义 2.3.5 设 H 为一个双代数, 称 $(M, \cdot, \rho)$ 为 H 上的一个左右**Long 模**, 若 $(M, \cdot)$ 为一个左 H-模, 而 (M, ρ) 为右 H-余模, 并且满足下面相容条件:

$$\rho(m \cdot h) = h \cdot m_0 \otimes m_{(1)}.$$

对于任意 $h \in H, m \in M$. 其中 $\rho(m) = m_0 \otimes m_{(1)}$.

由 H 上的左右 Long 模对象以及模同态和余模同态所组成的范畴记为 ${}_H\mathcal{L}^H$.

类似可以定义 H 上的右右 Long 模范畴 $\mathcal{L}_H^H$、左左 Long 模范畴 ${}_H^H\mathcal{L}$ 和右左 Long 模范畴 ${}^H\mathcal{L}_H$.

定义 2.3.6 已知 A 是一个双代数, 那么 A 上的一个**左积分**是一个元素 $\Lambda \in A$ 满足 $a\Lambda = \varepsilon(a)\Lambda$, 对任意 $a \in A$. 类似地, A 上的一个**右积分**是一个元素 $\Lambda \in A$ 满足 $\Lambda a = \Lambda \varepsilon(a)$, 对任意 $a \in A$.

2.3 积 分

我们将 A 上的全部左 (右) 积分构成的集合分别记为 $\int_l^H \left(\int_r^H \right)$, 易见它们都是 A 的理想.

更一般地, 有下面定义.

定义 2.3.7 设 H 是一个 Hopf 代数, A 是一个右 (左)H-余模代数. 一个右 (左)**积分**$\phi : H \longrightarrow A$ 是一个右 (左)H-余模映射, 即满足:

$$\sum \phi(h_1) \otimes h_2 = \sum \phi(h)_0 \otimes \phi(h)_{(1)} \quad \left(\sum h_1 \otimes \phi(h_2) = \sum \phi(h)_{(-1)} \otimes \phi(h)_0 \right)$$

对任意 $h \in H$. 进一步, ϕ 称为是右 (左)**全积分**(total integral)[62] 如果 $\phi(1_H) = 1_A$.

如果 $A = k$, 有 H^* 中的经典的右 (左) 积分:

$$\sum \phi(h_1)h_2 = \phi(h)1_H \quad \left(\sum h_1\phi(h_2) = \phi(h)1_H \right).$$

这等价于

$$\phi * h^* = h^*(1_H)\phi \quad (h^* * \phi = h^*(1_H)\phi).$$

定理 2.3.8 (i) 一个 Hopf 代数 H 是可分的当且仅当存在左或右积分 $t \in H$ 使得 $\varepsilon(t) = 1$.

(ii) H 是一个有限维半单的 Hopf 代数当且仅当存在左或右积分 $t \in H$ 使得 $\varepsilon(t) = 1$.

(iii) 每个有限维 Hopf 代数的反对极是双射.

定义 2.3.9 一个**幺模 Hopf 代数**(unimodular Hopf algebra) 是一个有限维 Hopf 代数使得 $\int_l \neq \int_r$.

关于积分与半单性质, 我们有如下定理.

定理 2.3.10 已知 H 是 Hopf 代数, 那么下面的性质是等价的:

(i) 所有的左 H-模完全可约.

(ii) H 是有限维的, $\varepsilon \left(\int_l \right) \neq (0)$.

(iii) H 是有限维的, 并且 $\varepsilon \left(\int_r \right) \neq (0)$.

(iv) 所有的 H-模是完全可约的.

(v) H 是一个幺模 Hopf 代数.

推论 2.3.11 已知 H 是一个有限维的 Hopf 代数, 那么:

(i) 若 H 是半单的, 则 H 是一个幺模 Hopf 代数.

(ii) H 是半单的当且仅当 $\varepsilon(\Lambda) \neq = 0$, Λ 是 H 的某些左的或者右的积分.

关于积分与对极, 我们主要有如下的性质.

定理 2.3.12 已知 H 是一个有限维 Hopf 代数, S 是它的对极, 那么:

(i) $S\left(\int_l\right) = \int_r$;

(ii) $S\left(\int_r\right) = \int_l$.

一个四角 B-Hopf 双模 (M, ρ^l, ρ^r) 首先, 由 S. Woronowicz 考虑, 随后, W. Nichols 证明张量积取在 B 上, 则四角 B-Hopf 双模范畴 ${}_B^B\mathcal{H}_B^B$ 是一个张量范畴 [166, 176, 263, 264].

定理 2.3.13 设 B 是一个 Hopf 代数, 那么四角 B-Hopf 双模范畴 ${}_B^B\mathcal{H}_B^B$ 是一个辫子张量范畴. 详细地, 设 $M, N, P \in {}_B^B\mathcal{H}_B^B$, 存在唯一一个 B-双模同态 $\sigma_{M,N}: M \otimes_B N \longrightarrow N \otimes_B M$, 使得对任意 $x \in M_l^{coB}, y \in M_r^{coB}$, 有 $\sigma_{M,N}(x \otimes y) = y \otimes x$. 进一步, $\sigma_{M,N}$ 是一个可逆的双余模同态, 且满足下面辫子方程

$$(\mathrm{id}_P \otimes \sigma_{M,N})(\sigma_{M,P} \otimes \mathrm{id}_N)(\mathrm{id}_M \otimes \sigma_{N,P})$$
$$=(\sigma_{N,P} \otimes \mathrm{id}_M)(\mathrm{id}_N \otimes \sigma_{M,P})(\sigma_{M,N} \otimes \mathrm{id}_P).$$

2.4 对 偶

这一节, 我们主要研究代数的线性对偶问题.

对任意 K-空间 V, 设 $V^* = \mathrm{Hom}(V, K)$ 表示 V 的**线性对偶**. V 和 V^* 通过公式 $\langle f, v \rangle = f(v)$ 决定了一个非退化的双线性形式 (bilinear form)$\langle,\rangle : V^* \otimes V \to K$. 如果 $\phi : V \to W$ 是 K-线性的, 那么 ϕ 的转置 (transpose) 是 $\phi^* : W^* \to v^*$, 即对任意 $f \in V^*, v \in V$, 有

$$\phi^*(f)(v) = f(\phi(v)). \tag{2.4.1}$$

现在从 2.1 节中弱余环的对偶研究起. 对任意一个弱 A-余环, 从 C 到 A 的所有 A-模映射集合有环结构. 首先有如下经典同构:

$$C^* := \mathrm{Hom}_{-A}(C, A) \simeq \mathrm{Hom}_{-A}(CA, A),$$

$$(AC)^* := \mathrm{Hom}_{-A}(AC, A) \simeq \mathrm{Hom}_{-A}(ACA, A),$$

$${}^*C := \mathrm{Hom}_{A-}(C, A) \simeq \mathrm{Hom}_{A-}(AC, A),$$

$${}^*(CA) := \mathrm{Hom}_{A-}(CA, A) \simeq \mathrm{Hom}_{A-}(ACA, A),$$

$${}^*C^* := \mathrm{Hom}_{AA}(C, A) \simeq \mathrm{Hom}_{AA}(ACA, A) \simeq {}^*C \cap C^*.$$

2.4 对偶

命题 2.4.1 具有上述记号, 则

(i) C^* 的环结构为: 对任意 $f,g \in C^*$

$$f *_r g : C \xrightarrow{\Delta} CA \otimes_A C \xrightarrow{g \otimes \mathbf{id}} A \otimes_A C \simeq AC \xrightarrow{f} A,$$

即 $(f *_r g)(c) = \sum f(g(c_1)c_2)$, ε 是 C^* 中的中心幂等元, 且 $(AC)^* = \varepsilon *_r C^*$.

(ii) $^*C$ 的环结构为: 对任意 $f,g \in C^*$

$$f *_l g : C \xrightarrow{\Delta} C \otimes_A AC \xrightarrow{\mathbf{id} \otimes f} C \otimes_A A \simeq CA \xrightarrow{g} A,$$

即 $(f *_l g)(c) = \sum g(c_1 f(c_2))$, ε 是 $^*C$ 中的中心幂等元, 且 $(CA)^* = \varepsilon *_l C^*$.

(iii) $^*C^*$ 的环结构为: 对任意 $f,g \in C^*$

$$f * g : C \xrightarrow{\Delta} C \otimes_A A \otimes_A C \xrightarrow{g \otimes \mathbf{id} \otimes f} A,$$

即 $(f * g)(c) = \sum g(c_1)f(c_2)$, 单位元 ε.

(iv) 如果 C 是余结合 R-余环, 那么环结构也是结合的.

注记 2.4.2 (i) 如果 C 是 R-余环, 那么 C^*, $^*C$ 和 $^*C^*$ 有单位元 ε.

(ii) 如果 (C, Δ, ε) 是一个余代数, 那么 $(^*C = C^*, m = \Delta^*, u = \varepsilon^*)$ 是一个代数. 如果 C 是余交换的, 那么 C^* 是交换的.

如果我们先给定一个任意代数 A, 取其对偶, 我们会遇到困难. 因为, 如果 A 不是有限维的, 则 $A^* \otimes A^*$ 是 $(A \otimes A)^*$ 的真子空间, 所以 $m^* : A^* \to (A \otimes A)^*$ 的像不一定在 $A^* \otimes A^*$ 中. 当然, 如果 A 是有限维的, 则 A^* 是一个余代数. 通常地, 我们有如下的定义.

定义 2.4.3 设 A 是一个 K-代数. A 的**有限对偶**(finite dual) 是 $A^o = \{f \in A^* | f(I) = 0\}$, 其中 I 是 A 的理想, 并且 $\dim A/I < \infty$.

命题 2.4.4 如果 (A, m, u) 是一个 K-代数, 则 $(A^o, \Delta = m^*, \varepsilon = u^*)$ 是一个余代数. 如果 A 是交换的, 那么 A^o 是余交换的.

标注 2.4.5 特别地, A^o 是 A^* 的使得 $m^*(V) \subseteq V \otimes V$ 成立的最大子空间 V.

命题 2.4.6 设 (A, m, u) 是一个代数, 并且 $F \in A^*$, 则下面的表述是等价的:

(i) $f(I) = 0$, 其中 I 是 A 的余有限右理想.

(ii) $f(I) = 0$, 其中 I 是 A 的余有限左理想.

(iii) $f(I) = 0$, 其中 I 是 A 的余有限理想.

(iv) $\dim(A \rightharpoonup f) < \infty$.

(v) $\dim(f \rightharpoonup A) < \infty$.

(vi) $\dim(A \rightharpoonup f \leftharpoonup A) < \infty$.

(vii) $m^* f \in A^* \otimes A^*$.

如果 (i)—(vii) 有一个成立, 则 $f \in A^o$.

命题 2.4.7 设 $(B, m, u, \Delta, \varepsilon)$ 是一个双代数, 则 $(B^o, \Delta^*, \varepsilon^*, m^*, u^*)$ 也是一个双代数.

例 2.4.8 如果 B 是一个双代数, 那么 B^o 是一个双代数. 特别地, 考虑 $B = KG$. 在这种情况下 B^o 称为群 G 上的**表示函数集**(the set of representative functions)$R_K(G)$, 即

$$B^o = R_K(G) = \{f \in (KG)^* | \dim_K \mathrm{span}\{G \cdot f\} < \infty\},$$

其中, 对任意 $x, y \in G, f \in (KG)^*, G$ 在 $(KG)^*$ 上的作用为 $(x \cdot f)(y) = f(yx)$. 对任意 $x \in G, f, g \in B^*, B^o$ 的代数结构为

$$(fg)(x) = \Delta^*(f \otimes g)(x) = (f \otimes g)(x \otimes x) = f(x)g(x)$$

对任意 $x, y \in B, f \in B^o, B^o$ 的余代数结构为

$$\Delta f(x \otimes y) = m^* f(x \otimes y) = f(xy),$$

但是作为 $B^o \otimes B^o$ 的一个元素, 我们并没有给出 Δf 的具体计算公式. 当 $|G| < \infty$ 时, 我们可以如下描述.

设 $\{p_x | x \in G\}$ 是 $(KG)^*$ 的一组基, 即对任意 $x, y \in G$, 有 $p_x(y) = \delta_{x,y}$, 则

$$\Delta p_x = \Sigma_{uv=x} p_u \otimes p_v.$$

例 2.4.9 (分次对偶) 设 $V = \oplus_{n \geq 0} V_n$ 是一个分次向量空间, 则它的分次对偶向量空间 $V_{\mathrm{gr}}^* = \oplus_{n \geq 0} V_n^*$ 是一个分次向量空间. 设 $W = \oplus_{n \geq 0} W_n$ 是另外一个分次向量空间, 则在 $V \otimes W$ 上存在一个分次使得

$$(V \otimes W)_n = \oplus_{i+j=n} V_i \otimes W_j.$$

进一步地, 如果对任意 $n \geq 0, V_n$ 都是有限维的, 那么有 $V_{\mathrm{gr}}^* \otimes W_{\mathrm{gr}}^* = (V \otimes W)_{\mathrm{gr}}^*$.

给定一个 H-模代数 A, 通过冲 (smash) 积[133] 我们可以构造一个新的代数 $A \sharp H$. 这个新代数是一个 H^o-模代数. 设 H^o 平凡的作用在 A 上并且通过通常的作用 $\rightharpoonup$ 作用在 H 上, 有如下命题.

命题 2.4.10 $(A \sharp H) \sharp H^o \cong A \otimes \mathcal{L}$, 其中 $\mathcal{L}$ 是 $\mathrm{End}_K(H)$ 的一个最大 (最稠密) 的子环.

命题 2.4.11 设 H 是一个有限维 Hopf 代数, H^* 通过 $\rightharpoonup$ 作用在 H 上, 那么作为代数, $H \sharp H^* \cong \mathrm{End}_K(H)$.

命题 2.4.12 (i) 如果 M 是一个右 C-余模, 则 M 是一个左 C^*-模.

(ii) 设 M 是一个左 A-模, 那么 M 是一个右 A^o-余模 $\Leftrightarrow$ 对任意 $m \in M$, $A \cdot m$ 是有限维的.

例 2.4.13 对任意余代数 C, 如果 $\rho = \Delta$, 则 $M = C$ 是一个右 C-余模. 由命题 2.4.12(i), 这就决定了一个 C 上的左 C^* 作用, 即对任意 $f \in C^*, c \in C$,

$$f \rightharpoonup c = \Sigma \langle f, c_2 \rangle c_1,$$

这个作用也可以表示成, 对任意 $f, g \in C^*, c \in C$,

$$\langle g, f \rightharpoonup c = \Sigma \langle f, c_2 \rangle \langle g, c_1 \rangle = (g * f)(c) = \langle gf, c \rangle.$$

设 G 是一个有限群, $A = \oplus_{x \in G} A_x$ 是一个分次代数, 秩为 n, 并且是一个 $(KG)^*$-模代数.

命题 2.4.14 我们有: $(A \sharp (KG)^*) \sharp KG \cong A \otimes M_n(K) \cong M_n(A)$.

设 $\{p_x | x \in G\}$ 是 $(KG)^*$ 的一组基, G 在 $(KG)^*$ 上的左、右作用分别为

$$y \rightharpoonup p_x = p_{xy^{-1}}, \quad p_x \leftharpoonup y = p_{y^{-1}x}$$

对任意的 $x, y \in G$. 因此, G 在 $A \sharp (KG)^*$ 上的左作用为

$$y \cdot (a \sharp p_x) = a \sharp (y \rightharpoonup p_x) = a \sharp p_{xy^{-1}}.$$

定义 2.4.15 I 是 A 的**分次理想**(graded ideal), 如果 $I = \oplus_x I_x$, 其中 $I_x = I \cap A_x$.

命题 2.4.16 (i) 对任意的 $A \sharp (KG)^*$ 理想 $\mathcal{I}$, $I = \mathcal{I} \cap A$ 是 A 的分次理想.

(ii) 如果 $\mathcal{I}$ 是 G-稳定的, $\mathcal{I} = (\mathcal{I} \cap A) \sharp (KG)^*$.

(iii) 如果 A 是分次半单的, 则 $A \sharp (KG)^*$ 是半单的.

(iv) 如果 P 是 A 的素理想, 则存在 $A \sharp (KG)^*$ 的一个素理想 $\mathcal{P}$ 使得 $\mathcal{P} \cap A = P_G$. $\mathcal{P}$ 在它的 G-轨道 $\{\mathcal{P}^x | x \in G\}$ 下是唯一的, 并且 $P_G \sharp (KG)^* = \cap_x \mathcal{P}^x$.

(v) 如果 $\mathcal{P}$ 是 $A \sharp (KG)^*$ 的任意素理想, 则对 A 的一些素理想 P 有 $\mathcal{P} \cap A = P_G$.

命题 2.4.17 设 Q 是 A 的一个分次素理想, 如果 $P \subset Q, P \neq Q$ 都是 A 的素理想, 则 $P \cap A_1 \neq Q \cap A_1$.

我们用如下的一些结论结束本节.

命题 2.4.18 (i) 设 (H, S) 是一个有限维 Hopf 代数, 那么双代数 (H^*, S^*) 是一个 Hopf 代数.

(ii) 一个有限维有色 Hopf 代数 H 的对偶仍然是一个有色 Hopf 代数.

(iii) 设 H 是一个 $\deg d = 1$ 的微分 Hopf 代数, 则 H^* 也是一个微分 Hopf 代数, 并且, 对任意 $x \in H, a \in H^*$ 它的微分 ∂ 具有次数 -1, 定义如下

$$\langle x, \partial a \rangle = \langle dx, a \rangle.$$

命题 2.4.19 设 (H,S) 是一个有限维连通的 Hopf 代数, 则下面的表述是等价的:

(i) H 是半单的.

(ii) H 是交换的并且是半单的.

(iii) $P(H)$ 是一个环面.

(iv) 对一些 p-群 G, $H^* \simeq K[G]$.

2.5 一类诱导的群胚

本节介绍 Yetter-Drinfel'd 模范畴以及相关的性质. 尤其给出一类相关的群胚结构, 主要涉及 Nichols 代数. Nichols 代数是一个辫子张量范畴中的 Hopf 代数, 它包含着某些泛性质且自然出现在每个有点 Hopf 代数中. 最有名的例子是小量子群的 Borel 部分 $u_q(\mathfrak{g})^{\pm}$. 事实上, 一般它也保留着 Lie-理论的成分, 且任何 Nichols 代数由一个 Weyl 群胚所控制 [10].

定义 2.5.1 设 H 为 Hopf 代数且其对极 S 为反对极. 一个左左**Yetter-Drinfel'd 模** V 是一个向量空间 V 带有一个左 H-作用 $\cdot$ 和一个左 H-余作用 ρ_l 并且满足下面条件:

$$\rho_l(h\cdot v) = h_1 v_{(-1)} S(h_3) \otimes (h_2 \cdot v_0), \quad h \in H, \quad v \in V. \tag{2.5.1}$$

这里, 记 $\rho_l(v) = v_{(-1)} \otimes v_0$, 对任意 $v \in V$.

所有左左 Yetter-Drinfel'd 模和左 H-线性左 H-余线性映射构成的范畴记为 $^H_H\mathcal{YD}$. 这个范畴 $^H_H\mathcal{YD}$ 是一个辫子张量范畴, 其辫子映射为

$$\tau: M \otimes N \longrightarrow N \otimes M, \quad m \otimes n \mapsto m_{(-1)} \cdot n \otimes m_0$$

对任意 $m, n \in M \in {}^H_H\mathcal{YD}$. 如 S 是双射, 那么 τ 是双射.

注记 2.5.2 (i) 公式 (2.5.1) 等价于下面条件:

$$\rho_l(h_1 \cdot v)(h_2 \otimes 1) = \Delta(h)\rho(v), \quad h \in H, \quad v \in V.$$

(ii) 类似地, 我们也有左右 Yetter-Drinfel'd 模范畴 $_H\mathcal{YD}^H$、右左 Yetter-Drinfel'd 模范畴 $^H\mathcal{YD}_H$ 和右右 Yetter-Drinfel'd 模范畴 $\mathcal{YD}^H_H$. 例如, 设 $M \in {}_H\mathcal{YD}^H$, 有

$$h_1 \cdot m_0 \otimes h_2 m_{(1)} = (h_2 \cdot m)_0 \otimes (h_2 \cdot m)_{(1)} h_1 \iff \rho_r(h \cdot m) = \Delta(h_2)\rho_r(m)(1 \otimes S^{-1}(h_1))$$

对任意 $h \in H, m \in M$.

例 2.5.3 设 G 是一个群, (V, ρ_V) 是左 kG-模. 对任意 $g \in G, v \in V$, 设 $v_g \in V$ 使得 $\rho_V(v) = \sum_{g \in G} g \otimes v_g$ (有限和), 那么由余结合性, 有

$$\sum_{g \in G} g \otimes g \otimes v_g = \sum_{g \in G} \Delta(g) \otimes v_g = \sum_{g \in G} g \otimes \rho_V(v_g),$$

那么 $\rho_V(v_g) = g \otimes v_g$ 对任意 $g \in G, v \in V$.

另一方面, $v = \sum_{g \in G} \varepsilon(g) v_g = \sum_{g \in G} v_g$. 因此, $V = \bigoplus_{g \in G} V_g$, 这里 $V_g = \{v \in V \mid \rho_V(v) = g \otimes v\}$.

假设 G 是交换群, $(V, \cdot, \rho_V) \in {}^{kG}_{kG}\mathcal{YD}$, 也称 $(V, \cdot, \rho_V)$ 是 G 上的 Yetter-Drinfel'd 范畴. 那么我们有 $\rho_V(h \cdot v) = hgh^{-1} \otimes h \cdot v = g \otimes h \cdot v$ 对任意 $g, h \in G, v \in V_g$, 因此, V_g 是 G-模. 假设 G 由特征作用在 $V_g, g \in G$ 上, 即 $V = \bigoplus_{g \in G, \xi \in \hat{G}} V_g^\xi$, 这里

$$V_g^\xi = \{v \in V_g \mid h \cdot v = \xi(h) v, \forall h \in G\},$$

我们说 V 是**对角型**.

注记 2.5.4 如果 k 是特征为零的代数闭域, 且 G 是有限交换群, 那么 G 上所有有限维 Yetter-Drinfel'd 模都是对角型的.

命题 2.5.5 设 H 是 Hopf 代数, $V, W \in {}^H_H\mathcal{YD}$, 那么 $V \otimes W \in {}^H_H\mathcal{YD}$, 这里

$$\rho(v \otimes w) = v_{(-1)} w_{(-1)} \otimes (v_0 \otimes w_0), \quad h \cdot (v \otimes w) = h_1 \cdot v \otimes h_2 \cdot w$$

对任意 $v \in V, w \in W, h \in H$. 如果 H 有双射反对极且 $\dim V < 1$, 那么 $V^* \in {}^H_H\mathcal{YD}$, 这里

$$(v^*)_{(-1)}(v^*)_0(v) = S^{-1}(v_{(-1)}) v^*(v_0), \quad (h \cdot v^*)(v) = v^*(S(h) \cdot v)$$

对任意 $v \in V, v^* \in V^*, h \in H$.

命题 2.5.6 一个 Yetter-Drinfel'd 模范畴是 Grothendieck 范畴, 具有足够的内射对象.

定义 2.5.7 一个**辫子双代数** $(R, m, u, \Delta, \varepsilon) \in {}^H_H\mathcal{YD}$ 是指

(i) (R, m, u) 是 ${}^H_H\mathcal{YD}$ 中的代数, 于是 $R \underline{\otimes} R$ 是 ${}^H_H\mathcal{YD}$ 中的代数, 具有乘法:

$$(a \otimes b)(c \otimes d) = a(b_{(-1)} \cdot c) \otimes b_0 d, \quad \forall a, b, c, d \in R.$$

(ii) (R, Δ, ε) 是 ${}^H_H\mathcal{YD}$ 中的余代数.

(iii) $\Delta : R \to R \underline{\otimes} R$ 和 $\varepsilon : R \to K$ 是代数同态.

如果 id_H 卷积可逆于 S_H, 那么称 H 是 ${}^H_H\mathcal{YD}$ 中的**辫子 Hopf 代数**(braided Hopf algebra).

例 2.5.8 (1) 设 H 是一个 Hopf 代数, 且 $V \in {}_H^H\mathcal{YD}$. 那么张量代数 $T(V) = \bigoplus_{n=0}^{\infty} V^{\otimes n}$ 是一个辫子 Hopf 代数, 这里

$$\Delta(v) = 1 \otimes v + v \otimes 1 \in T(V) \otimes T(V), \quad \varepsilon(v) = 0, \quad \forall v \in V.$$

例如, 设 $u, v, w \in V$. 简单地写 $T(V)$ 中的乘法不带张量积符号, 那么有

$$\begin{aligned}\Delta(u + vw) &= \Delta(u) + \Delta(v)\Delta(w) \\ &= 1 \otimes u + u \otimes 1 + (1 \otimes v + v \otimes 1)(1 \otimes w + w \otimes 1) \\ &= 1 \otimes u + u \otimes 1 + 1 \otimes vw + (v_{(-1)} \cdot w) \otimes v_0 + v \otimes w + vw \otimes 1.\end{aligned}$$

如果 $H = kG$ 是群代数, V 是对角型的, $g \in G, v \in V_g, w \in V$, 那么

$$\Delta(vw) = 1 \otimes vw + (g \cdot w) \otimes v + v \otimes w + vw \otimes 1.$$

对于一个辫子 Hopf 代数 B, 我们写 $\Delta(b) = b^1 \otimes b^2 \in B \otimes B$.

(2) 设 H 是平凡群的群代数, 即 $H = k1$. 那么 B 是 ${}_H^H\mathcal{YD}$ 中辫子 Hopf 代数当且仅当 B 是一般的 Hopf 代数.

定理 2.5.9 (Radford 双积 [155], Majid 玻色化 (Bosonization)[126, 128]) 设 H 是 Hopf 代数, B 是 ${}_H^H\mathcal{YD}$ 中辫子 Hopf 代数. 那么 $B \# H = B \otimes H$(作为向量空间) 上可以定义乘积 (冲积) 如下:

$$(b\#h)(a\#l) = b(h_1 \cdot a)\#h_2 l, \quad a, b \in B, h, l \in H$$

和余乘积 (冲余积):

$$\Delta(b\#h) = (b^1 \otimes b^2{}_{(-1)} h_1) \otimes (b^2{}_0 \otimes h_2), \quad b \in B, h \in H.$$

那么, H 是一个 Hopf 代数具有反对极:

$$S(b\#h) = (1 \otimes S(h)S(b_{(-1)}))(S_B(b_0) \otimes 1), \quad b \in B.$$

注记 2.5.10 (i) 玻色化是 N.Andruskiewitsch 和 H. J. Schneider 提升方法的起点 [12, 13], 这种方法可以分类有点 Hopf 代数 (即所有单子余代数是一维的) 或更一般 Hopf 代数, 它的余根是一个 Hopf 子代数.

主要原理是: 已知一个非余半单 (noncosemisimple)Hopf 代数 A(例如有点 Hopf 代数), 考虑它的余根滤: $A_0 \subseteq A_1 \subseteq \cdots \subseteq A$, 这是一个余代数滤. 再考虑结合的分次余代数 $\mathrm{gr}(A) = \bigoplus_{n \geqslant 0} \mathrm{gr}(A)(n), \mathrm{gr}(A)(n) = A_n/A_{n-1}$, 这里 $A_{-1} = 0$. 因为 A 的余根 A_0 是一个 Hopf 子代数, 所以 $\mathrm{gr}(A)$ 是一个分次 Hopf 代数且它自己的余根

滤的零项 $\mathrm{gr}(A)(0) = A_0$ 是 $\mathrm{gr}(A)$ 的 Hopf 子代数. 令 $B = \mathrm{gr}(A), H = \mathrm{gr}(A)(0)$. 设 $\gamma: H \longrightarrow B$ 是包含映射和 $\pi: B \longrightarrow H$ 是投射, 带有核 $\mathrm{Ker}(\pi) = \bigoplus_{n \geq 1} \mathrm{gr}(A)(n)$. 设
$$R = B^{\mathrm{coH}} = \{a \in B \mid (\mathbf{id} \otimes \pi)\Delta(a) = a \otimes 1\},$$
那么 π 是 γ 的一个 Hopf 代数收缩 (retraction) (或 γ 是 π 的一个截面), 即 $\pi \circ \gamma = \mathbf{id}_H$. 于是, 由 Radford 和 Majid 结论可知, $R \in {}_H^H\mathcal{YD}$ 是一个辫子 Hopf 代数.

详细的结构如下: H 在 R 上的作用是通过伴随表示与 γ 的合成给出. 余作用是 $(\pi \otimes \mathbf{id})\Delta$. 这两个结构通过 Yetter-Drinfel'd 条件相关联:
$$\rho_R(h \cdot r) = h_1 r_{(-1)} S(h_3) \otimes h_2 \cdot r_0$$
对任意 $r \in R, h \in H$. 因此, R 是 ${}_H^H\mathcal{YD}$ 中一个对象.

进一步, R 是 B 一个子代数且是一个余代数: 余乘法为
$$\Delta_R(r) = r_1 \gamma \pi S(r_2) \otimes r_3 =: \sum r^{(1)} \otimes r^{(2)};$$
余单位是 B 的余单位在 R 上的限制. 不难验证 $\Delta_R m = (m \otimes m)(\mathbf{id} \otimes \tau \otimes \mathbf{id})(\Delta_R \otimes \Delta_R)$, 这里 m 表示 R 的乘法. R 上的反对极 $S_R: R \longrightarrow R, r \mapsto \gamma\pi(r_1) S_B(r_2)$. 进一步有: $B = R \# H$.

那么要研究 A, 首先研究 R, 然后通过玻色过程传递信息给 B, 最后通过过滤提升到 A.

(ii) 设 K 为域, H 是 K 上任意 Hopf 代数. 令 $g \in G(H), \chi \in \mathrm{Alg}(H, K)$ 使得
$$\chi(h)g = h_1 \chi(h_2) g S(h_3), \quad h \in H.$$
设 N 是 $q := \chi(g)$ 的阶, 假设 N 是有限的.

设 $R = K[y]/(y^N)$, 那么 $R \in {}_H^H\mathcal{YD}$ 是一个辫子 Hopf 代数, 具有 H-模结构和 H-余模结构为
$$h \cdot y^t = \chi^t(h) y^t, \quad \rho_R(y^t) = g^t \otimes y^t,$$
余乘为 $\Delta_R(y) = y \otimes 1 + 1 \otimes y$. 该辫子 Hopf 代数记为 $\mathbb{R}(g, \chi)$. 我们有: $\mathbb{R}(g, \chi) \cong \mathbb{R}(g', \chi')$ 如果 $g = g', \chi = \chi'$.

下面命题很有用.

命题 2.5.11 设 $B \in {}_H^H\mathcal{YD}$ 是一个辫子 Hopf 代数. 设 $B' \subset B, B' \in {}_H^H\mathcal{YD}$ 是一个子代数且 $I \subset B \cap \mathrm{Ker}\varepsilon, I \in {}_H^H\mathcal{YD}$ 是 B' 的一个理想. 如果
$$\Delta(B') \subset B' \otimes B' + B \otimes I, \quad \Delta(I) \subset I \otimes B' + B \otimes I.$$
那么 B 上的辫子 Hopf 代数结构诱导出 $B'/I \in {}_H^H\mathcal{YD}$ 上的辫子 Hopf 代数结构.

设 H 是 Hopf 代数且 $V \in {}_H^H\mathcal{YD}$. 回顾

$$T(V) = \bigoplus_{n=0}^{\infty} V^{\otimes n}$$

是一个 $\mathbb{N}_0$-分次代数, 这里 $\deg x = n$ 对 $x \in V^{\otimes n}, n \in \mathbb{N}_0$.

引理 2.5.12　在 $T(V)$ 中, 存在唯一一个最大余理想 $\mathfrak{J}(V)$ 包含在 $T^{++}(V) = \bigoplus_{n \geqslant 2} V^{\otimes n}$ 中. 余理想 $\mathfrak{J}(V)$ 关于 $T(V)$ 的 $\mathbb{N}_0$-分次是齐次的.

证明　设 $(D_i)_{i \in I}$ 是一簇 $T(V)$ 的所有余理想包含在 $T^{++}(V)$ 中, 那么 $D = \sum_{i \in I} D_i \subset T^{++}(V)$ 一个 $T(V)$ 的余理想. 事实上, 对任意 $n \in \mathbb{N}, x_i \in D_i, i \in \{i_1, i_2, \cdots, i_n\} \subset I$. 那么,

$$\Delta\left(\sum_{j=1}^n x_{i_j}\right) = \sum_{j=1}^n \Delta(x_{i_j}) \in \sum_{j=1}^n (D_{i_j} \otimes T(V) + T(V) \otimes D_{i_j})$$
$$\subset D \otimes T(V) + T(V) \otimes D.$$

因此, $D = D_{i_0}$ 对某个 $i_0 \in I$ 和 $D_i \subset D$ 对任意 $i \in I$. 我们得到 $\mathfrak{J}(V) = D$.

下面证明余理想 $\mathfrak{J}(V)$ 关于 $T(V)$ 的 $\mathbb{N}_0$-分次是齐次的. 设 $\pi_n : T(V) \longrightarrow V^{\otimes n}$ 是自然投射. 定义 $D = \bigoplus_{n \geqslant 2} \pi_n(\mathfrak{J}(V))$. 那么 $D \subset T^{++}(V)$. 因为 $\Delta(1) = 1 \otimes 1$, $\Delta(v) = v \otimes 1 + 1 \otimes v$ 对任意 $v \in V$. 对 $n = 0, 1$, 有

$$\Delta(x) = \bigoplus_{m=0}^n \pi_m(x_1) \otimes \pi_{n-m}(x_2), \quad 对任意 x \in V^{\otimes n}.$$

定义 2.5.13　商余代数 $\mathfrak{B}(V) = T(V)/\mathfrak{J}(V)$ 称为是 V 的 **Nichols 代数**.

定理 2.5.14　商余代数 $\mathfrak{B}(V) = T(V)/\mathfrak{J}(V)$ 是 $\mathbb{N}_0$-分次辫子 Hopf 代数.

一般地, 一个 ${}_H^H\mathcal{YD}$ 中的**分次辫子 Hopf 代数**(graded braided Hopf algebra) 是一个 ${}_H^H\mathcal{YD}$ 中的**辫子 Hopf 代数**, 并且 $R = \oplus_{n \geqslant 0} R(n), R(n) \in {}_H^H\mathcal{YD}$ 使得 H 是一个分次代数和一个分次余代数.

设 R 是一个 ${}_H^H\mathcal{YD}$ 中的有限维辫子 Hopf 代数, 则对偶空间 $S = R^*$ 是一个 ${}_H^H\mathcal{YD}$ 中辫子 Hopf 代数. 更进一步, 如果 $R = \oplus_{n \geqslant 0} R(n), R(n) \in {}_H^H\mathcal{YD}$ 是一个齐次项均为有限维的分次辫子 Hopf 代数, 则分次对偶 $S = R^* = \oplus_{n \geqslant 0} R(n)^*$ 是一个 ${}_H^H\mathcal{YD}$ 中的分次辫子 Hopf 代数.

设 $P(R)$ 表示 R 的本原元素集合, 下面给出一般定义.

定义 2.5.15　设 $V \in {}_H^H\mathcal{YD}$, 如果 $\mathbb{K} \simeq R(0), V \simeq R(1)$ 并且

$$P(R) = R(1), \tag{2.5.2}$$

R 作为一个代数是由 $R(1)$ 生成的, $\tag{2.5.3}$

则 $_H^H\mathcal{YD}$ 中的一个分次辫子 Hopf 代数 $R = \oplus_{n \geq 0} R(n)$ 称为 V 的 **Nichols 代数**(Nichols algebra).

V 的维数称为 R 的**秩**(rank). 由定理 2.5.14 可知, $V \in {}_H^H\mathcal{YD}$ 的 **Nichols 代数**存在并且是唯一的.

注记 2.5.16 (i) 张量代数 $T(V) = \oplus_{n \geq 0} T(V)(n)$ 是 $_H^H\mathcal{YD}$ 中的一个分次辫子 Hopf 代数.

(ii) 考虑集合 $\wp = \{I | I \subset T(V)\}$ 使得

(a) I 是由次数大于等于 2 的齐次项生成的齐次理想.

(b) I 是一个余理想.

这里我们并没有要求 $I \in {}_H^H\mathcal{YD}$. 设 $\tilde{\wp} = \{I | I \in \wp, I \in {}_H^H\mathcal{YD}\} \subset \wp$. 理想

$$I(V) = \sum_{I \in \wp} I, \quad \tilde{I}(V) = \sum_{J \in \tilde{\wp}} J$$

分别是 $\wp, \tilde{\wp}$ 中最大的元素.

命题 2.5.17 设 $\mathcal{B}(v) := T(v)/\tilde{I}(V)$, 则

(i) $V = P(\mathcal{B}(v))$, 因此 $\mathcal{B}(v)$ 是 V 的 Nichols 代数.

(ii) $I(V) = \tilde{I}(V)$.

(iii) 设 $H = \oplus_{n \geq 0} H(n)$ 是 $_H^H\mathcal{YD}$ 中的一个分次辫子 Hopf 代数, 使得 $R(0) = \mathbb{K}$ 并且 H 作为一个代数是由 $V := H(1)$ 生成的. 那么存在一个分次 Hopf 代数的满射 $H \to \mathcal{B}(v)$, 它是一个次数为 1 的 Yetter-Drinfel'd 模同构.

(iv) 设 $H = \oplus_{n \geq 0} H(n)$ 是 V 的 Nichols 代数, 则 $H \simeq \mathcal{B}(v)$ 是 $_H^H\mathcal{YD}$ 中的辫子 Hopf 代数.

(v) 设 $H = \oplus_{n \geq 0} H(n)$ 是 $_H^H\mathcal{YD}$ 中的一个分次辫子 Hopf 代数, 使得 $R(0) = \mathbb{K}$ 并且 $H(1) = P(H) = V$, 则 $\mathcal{B}(v)$ 与 H 的由 V 生成的子代数 $K\langle V\rangle$ 同构.

如果 U 是 $V \in {}_H^H\mathcal{YD}$ 的一个辫子子空间, 即 $c(U \otimes U) \subset U \otimes U$, c 是 V 的辫子, 我们定义 $\mathcal{B}(U) := T(U)/I(V)$, 则可得如下结论.

命题 2.5.18 (i) 分配 (assignment)$V \to \mathcal{B}(v)$ 是一个从范畴 $_H^H\mathcal{YD}$ 到 $_H^H\mathcal{YD}$ 中的辫子 Hopf 代数范畴的函子.

(ii) 如果 U 是 V 的子 Yetter-Drinfel'd 模, 或者更一般地, U 是 V 的辫子子空间, 则正则映射 $\mathcal{B}(U) \to \mathcal{B}(v)$ 是单射.

命题 2.5.19 设 $H = \oplus_{n \geq 0} H(n)$ 是 $_H^H\mathcal{YD}$ 中的一个分次辫子 Hopf 代数, 齐次项都是有限维的并且 $H(0) = \mathbb{K}1$. 设 $S = \oplus_{n \geq 0} H(n)^*$ 是 H 的分次对偶, 则 $H(1) = P(H)$ 当且仅当 S 作为代数是由 $S(1)$ 生成的.

例 2.5.20 设 F 是一个特征为正 p 的域, S 是通常的 Hopf 代数 $F[x]/\langle x^{p^2}\rangle$ 并且 $x \in P(S)$, 则 $x^p \in P(S)$. 因此, S 满足 (2.5.3) 但不满足 (2.5.1).

例 2.5.21 设 $S = \mathbb{K}[X] = \oplus_{n \geq 0} S(n)$ 是一个只有一个变量的多项式代数. 我们考虑将 S 作成 $^H_H \mathcal{YD}$ 中的辫子 Hopf 代数. 其中, $H = \mathbb{K}\Gamma$, Γ 是一个生成元为 g 的无穷循环群, 作用、余作用、余乘如下定义:

$$\rho(X^n) = g^n \otimes X^n, \quad g \cdot X = qX, \quad \Delta(X) = X \otimes 1 + 1 \otimes X,$$

其中, q 为 1 的 N 次单位根, 即 S 就是所谓的量子线 (quantum line). 则 S 满足 (2.5.2) 但不满足 (2.5.1), 因为 X^N 是素的. 从而, 分次对偶 $H = S^d = \oplus_{n \geq 0} S(n)^*$ 是一个辫子 Hopf 代数满足 (2.5.2) 但不满足 (2.5.3).

命题 2.5.22 $^H_H \mathcal{YD}$ 中的任意有限维辫子 Hopf 代数满足 (2.5.2) 和 (2.5.3).

第3章 弱 Hopf 代数

一般有限群上的函数代数是一个 Hopf 代数 [2], 如果有限群被替代为有限群胚, 那么其上的函数代数就是弱 Hopf 代数 (weak Hopf algebras). 弱 Hopf 代数是一般 Hopf 代数的推广, 与其他弱化形式的 Hopf 代数 (如拟 Hopf 代数和 Hom Hopf 代数等) 相比, 弱 Hopf 代数的特殊性在于它保持了 Hopf 代数的余结合性. 但它不再要求余乘运算保持单位和余单位运算是代数同态. 这就使得 Hopf 代数理论中许多重要性质在弱 Hopf 代数理论中都有其"弱"的形式. 本章主要介绍弱 Hopf 代数、积分、Drinfel'd 量子偶、深度 2 扩张以及对偶理论. 这些研究见文献 [9, 21, 39, 40, 74, 107, 110, 118, 140, 141, 167, 170, 224, 274, 279, 282].

3.1 弱 Hopf 代数定义

作为余环的推广, 我们讨论弱余环概念 [34, 261, 287].

设 A 是带有单位元 1_A 的结合代数. M 是非单位双 A-模, 即 $(am)b = a(mb)$, 但有 $1_A m \neq m \neq m 1_A$ 对任意 $a, b \in A, m \in M$. 显然, M 是单位双 A-模, 要求 $1_A m = m = m 1_A$. 用 $_A \widetilde{\mathbb{M}}_A$ 和 $_A \mathbb{M}_A$ 分别表示非单位双 A-模范畴和单位双 A-模范畴.

设 $M \in \widetilde{\mathbb{M}}_A$, 那么存在一个可裂 A-满同态

$$- \otimes 1_A : M \longrightarrow M \otimes_A A, \quad m \mapsto m \otimes 1_A,$$

这是单射 (双射) 当且仅当 M 是单位 A-模, 即 $m = m 1_A$.

有经典同构: $M \otimes_A A \longrightarrow MA, m \otimes a \mapsto ma$, $\mathrm{Hom}_{-A}(A, M) \longrightarrow MA, f \mapsto f(1_A)$, 我们经常把这些模视为相同, 尤其, $MA = M 1_A$.

我们有函子:

$$- \otimes_A A : \widetilde{\mathbb{M}}_A \longrightarrow \mathbb{M}_A \subset \widetilde{\mathbb{M}}_A, \quad M \mapsto M \otimes_A A, \quad f \mapsto f \otimes \mathrm{id}.$$

这个函子左 (右) 伴随自身, 即: $M, N \in \widetilde{\mathbb{M}}_A$, 有

$$\mathrm{Hom}_{-A}(M \otimes_A A, N) \simeq \mathrm{Hom}_{-A}(M \otimes_A A, N \otimes_A A) \simeq \mathrm{Hom}_{-A}(M, N \otimes_A A).$$

因为 A 是单位 A-模, 所以 $\mathrm{Hom}_{-A}(M, A) \simeq \mathrm{Hom}_{-A}(MA, A)$.

设 $M \in {}_A\widetilde{\mathbb{M}}_A$, 那么存在一个可裂 (A,A)-满同态

$$1_A \otimes - \otimes 1_A : M \longrightarrow A \otimes_A M \otimes_A A \simeq AMA, \quad m \mapsto 1_A \otimes m \otimes 1_A (= 1_A m 1_A)$$

和同构:

$$\mathrm{Hom}_{AA}(M,A) \simeq \mathrm{Hom}_{AA}(MA,A) \simeq \mathrm{Hom}_{AA}(AMA,A).$$

定义 3.1.1 设 $C \in {}_A\widetilde{\mathbb{M}}_A$.

(i) 一个双 (A,A)-双模同态 $\Delta : C \longrightarrow C \otimes_A A \otimes_A C$ 称为是**弱余乘法**(weak comultiplication). 对任意 $c \in C$, 记 $\Delta(c) = \sum c_1 \otimes 1_A \otimes c_2$. 弱余乘 Δ 称为是**余结合的**如果有

$$\sum c_{11} \otimes 1_A \otimes c_{12} \otimes 1_A \otimes c_2 = \sum c_1 \otimes 1_A \otimes c_{21} \otimes 1_A \otimes c_{22}$$

对任意 $c \in C$.

(ii) 一个双 (A,A)-双模同态 $\varepsilon : C \longrightarrow A$ 称为是关于 Δ 的**弱余单位元**(weak counit) 如果

$$\sum \varepsilon(c_1) c_2 = 1_A c 1_A = \sum c_1 \varepsilon(c_2)$$

对任意 $c \in C$.

(iii) C 称为是**弱余环**(weak coring)(即弱 A-余环) 如果它有一个弱余乘法和弱余单位元.

(iv) 一个弱 A-余环 C 称为是右 (左) 单位如果 C 是右 (左) 单位 A-模.

(v) 一个弱 A-余环 C 称为是 A-余环如果 C 是双单位 (A,A)-模, 此时, 我们有: $C \otimes_A A \otimes_A C \simeq C \otimes_A C$.

(vi) 一个弱 A-余环 C 称为是 A-**余代数**如果 A 是交换的且左 A-模和右 A-模相同, 即 $ac = ca$ 对任意 $a \in A, c \in C$.

命题 3.1.2 设 C 是弱 A-余环. 那么

(i) $(CA, \Delta, \varepsilon)$ 是一个右单位弱 A-余环.

(ii) $(AC, \Delta, \varepsilon)$ 是一个左单位弱 A-余环.

(iii) $(ACA, \Delta, \varepsilon)$ 是一个 A-余环.

下面我们给出弱 Hopf 代数的相关定义.

定义 3.1.3 若五元组 $(H, m, u, \Delta, \varepsilon)$ 满足

(i) (H, m, u) 为域 k 上结合代数. 记 $1 := u(1)$ 为单位元.

(ii) (H, Δ, ε) 为域 k 上余代数. 记 $\Delta(x) = x_1 \otimes x_2$.

(iii) 对任意 $x, y, z \in H$, 有

$$\Delta(xy) = \Delta(x) \Delta(y),$$

3.1 弱 Hopf 代数定义

$$\varepsilon(xyz) = \varepsilon(xy_1)\varepsilon(y_2z), \quad \varepsilon(xyz) = \varepsilon(xy_2)\varepsilon(y_1z),$$

$$\Delta^2(1) = (\Delta(1) \otimes 1)(1 \otimes \Delta(1)), \quad \Delta^2(1) = (1 \otimes \Delta(1))(\Delta(1) \otimes 1).$$

则称 H 为**弱双代数**, 若同时存在 k-线性映射 $S : H \longrightarrow H$(称为**对极**) 满足:

(iv)

$$x_1 S(x_2) = \varepsilon(1_1 x)1_2, \quad S(x_1)x_2 = 1_1\varepsilon(x1_2), \quad S(x_1)x_2S(x_3) = S(x).$$

则称弱双代数 H 为**弱 Hopf 代数**.

对任意的弱 Hopf 代数 H, 对极 S 既是反代数同态, 又是反余代数同态且有 $S(1) = 1, \varepsilon \circ S = \varepsilon$, 若 H 是有限维的, 则 S 是双射.

命题 3.1.4 (1) 设 H 是一个弱 Hopf 代数. 下列情况等价:

(i) H 是 Hopf 代数; (ii) $\Delta(1) = 1 \otimes 1$; (iii) $\varepsilon(xy) = \varepsilon(x)\varepsilon(y)$;
(iv) $S(x_1)x_2 = 1\varepsilon(x)$; (v) $x_1S(x_2) = 1\varepsilon(x)$.

(2) 设 H 是一个弱双代数. 对任意 $a \in A$, 下列各式等价:

(i) $\Delta(a) = a1_1 \otimes 1_2$; (ii) $\Delta(a) = 1_1 a \otimes 1_2$;
(iii) $a = \varepsilon(a1_1)1_2$; (iv) $a = \varepsilon(1_1 a)1_2$.

(3) 设 H 是一个弱双代数. 对任意 $a \in A$, 下列各式等价:

(i) $\Delta(a) = 1_1 \otimes 1_2 a$; (ii) $\Delta(a) = 1_1 \otimes a1_2$;
(iii) $a = 1_1\varepsilon(1_2 a)$; (iv) $a = 1_1\varepsilon(a1_2)$.

例 3.1.5 (1) **弱缠绕结构** 设 A 是代数, C 是余代数.

(i) **右-右弱缠绕结构** (A, C, α) 线性映射 $\alpha : A \otimes C \longrightarrow A \otimes C, a \otimes c \mapsto a_\alpha \otimes c^\alpha$ 满足, 对任意 $a, b, c \in H$

$$(ab)_\alpha \otimes c^\alpha = a_\alpha b_\beta \otimes c^{\alpha\beta}; \tag{3.1.1}$$

$$a_\alpha \varepsilon(c^\alpha) = \varepsilon(c^\alpha)1_\alpha a; \tag{3.1.2}$$

$$a_\alpha \otimes \Delta(c^\alpha) = a_{\alpha\beta} \otimes c_1{}^\beta \otimes c_2{}^\alpha; \tag{3.1.3}$$

$$1_\alpha \otimes c^\alpha = \varepsilon(c_1{}^\alpha)1_\alpha \otimes c_2. \tag{3.1.4}$$

设 A 是一个弱双代数, 线性映射 $\alpha : A \otimes A \longrightarrow A \otimes A$ 定义为 $\alpha(a \otimes b) = (1 \otimes b)\Delta(a)$. 容易验证 (A, A, α) 是一个右-右弱缠绕结构. 注意, (3.1.1) 用到 Δ 是代数同态和代数的结合性, (3.1.3) 既用到 Δ 是代数同态, 也用到 Δ 的余结合性. 比如: 对于 (3.1.3), 有

$$a_1 \otimes (ca_2)_1 \otimes (ca_2)_2 = a_{11} \otimes c_1 a_{12} \otimes c_2 a_2$$

对任意 $a, c \in H$.

(ii) **左-右弱缠绕结构** (A, C, α)　线性映射 $\alpha : A \otimes C \longrightarrow A \otimes C, a \otimes c \mapsto a_\alpha \otimes c^\alpha$ 满足, 对任意 $a, b, c \in H$

$$(ab)_\alpha \otimes c^\alpha = a_\alpha b_\beta \otimes c^{\beta\alpha}; \tag{3.1.5}$$

$$a_\alpha \varepsilon(c^\alpha) = \varepsilon(c^\alpha) a 1_\alpha; \tag{3.1.6}$$

$$a_\alpha \otimes \Delta(c^\alpha) = a_{\alpha\beta} \otimes c_1{}^\beta \otimes c_2{}^\alpha; \tag{3.1.7}$$

$$1_\alpha \otimes c^\alpha = \varepsilon(c_1{}^\alpha) 1_\alpha \otimes c_2. \tag{3.1.8}$$

设 A 是一个弱双代数, 线性映射 $\alpha : A \otimes A \longrightarrow A \otimes A$ 定义为 $\alpha(a \otimes b) = \Delta(a)(1 \otimes b)$. 容易验证 (A, A, α) 是一个左-右弱缠绕结构.

(iii) **右-左弱缠绕结构** (A, C, α)　线性映射 $\alpha : C \otimes A \longrightarrow C \otimes A, c \otimes a \mapsto c^\alpha \otimes a_\alpha$ 满足, 对任意 $a, b, c \in H$

$$c^\alpha \otimes (ab)_\alpha = c^{\alpha\beta} \otimes a_\alpha b_\beta; \tag{3.1.9}$$

$$a_\alpha \varepsilon(c^\alpha) = \varepsilon(c^\alpha) 1_\alpha a; \tag{3.1.10}$$

$$\Delta(c^\alpha) \otimes a_\alpha = c_1{}^\alpha \otimes c_2{}^\beta \otimes a_{\alpha\beta}; \tag{3.1.11}$$

$$c^\alpha \otimes 1_\alpha = c_1 \otimes \varepsilon(c_2{}^\alpha) 1_\alpha. \tag{3.1.12}$$

设 A 是一个弱双代数, 线性映射 $\alpha : A \otimes A \longrightarrow A \otimes A$ 定义为 $\alpha(a \otimes b) = (a \otimes 1)\Delta(b)$. 容易验证 (A, A, α) 是一个右-左弱缠绕结构.

(iv) **左-左弱缠绕结构** (A, C, α)　线性映射 $\alpha : C \otimes A \longrightarrow C \otimes A, c \otimes a \mapsto c^\alpha \otimes a_\alpha$ 满足, 对任意 $a, b, c \in H$

$$c^\alpha \otimes (ab)_\alpha = c^{\alpha\beta} \otimes a_\beta b_\alpha; \tag{3.1.13}$$

$$\varepsilon(c^\alpha) a_\alpha = \varepsilon(c^\alpha) a 1_\alpha; \tag{3.1.14}$$

$$\Delta(c^\alpha) \otimes a_\alpha = c_1{}^\alpha \otimes c_2{}^\beta \otimes a_{\alpha\beta}; \tag{3.1.15}$$

$$c^\alpha \otimes 1_\alpha = c_1 \otimes \varepsilon(c_2{}^\alpha) 1_\alpha. \tag{3.1.16}$$

设 A 是一个弱双代数, 线性映射 $\alpha : A \otimes A \longrightarrow A \otimes A$ 定义为 $\alpha(a \otimes b) = \Delta(b)(a \otimes 1)$. 容易验证 (A, A, α) 是一个左-左弱缠绕结构.

(2) $(B, m, 1_B, \Delta, \varepsilon)$ 设弱双代数. 考虑 $B \otimes B$ 为双 (B, B)-模如下: 对任意 $a, b, c \in B$

$$(a \otimes b) \cdot c = (a \otimes b)\Delta(c),$$
$$a \cdot (b \otimes c) = ab \otimes c,$$

3.1 弱 Hopf 代数定义

那么 $A = B \otimes B$ 有 B-余环结构如下:

$$\Delta : B \otimes B \longrightarrow (B \otimes B) \otimes_B (B \otimes B) \simeq (B \otimes B) \cdot 1 \otimes B,$$

$$\Delta(a \otimes b) = \sum (a \otimes b_1) \otimes_B (1 \otimes b_2) = \sum a1_1 \otimes b_1 1_2 \otimes b_2,$$

$$\varepsilon : B \otimes B \longrightarrow (D \otimes D) \cdot 1 \xrightarrow{\mathrm{id} \otimes \varepsilon} B,$$

$$\varepsilon : a \otimes b \mapsto (a \otimes b) \cdot 1 \mapsto \sum a1_1 \varepsilon(b1_2).$$

设 $\varepsilon^{\mathrm{cop}}(a \otimes b) = \sum a1_2 \varepsilon(b1_1)$.

命题 3.1.6 (1) 设 A 是一个具有单位元的非结合代数且是具有余单位的非余结合余代数. 定义线性映射 $\alpha : A \otimes A \longrightarrow A \otimes A$ 为 $\alpha(a \otimes b) = \Delta(a)(1 \otimes b)$ 和 $\beta : A \otimes A \longrightarrow A \otimes A$ 定义为 $\beta(a \otimes b) = (a \otimes 1)\Delta(b)$. 那么 A 是一个弱双代数当且仅当 (A, A, α) 是一个左-右弱缠绕结构, 且 (A, A, β) 是一个右-左弱缠绕结构. 更进一步, A 是一个弱 Hopf 代数当且仅当 $\alpha : A \otimes A \longrightarrow A \otimes A$, $\beta : A \otimes A \longrightarrow A \otimes A$ 是双射.

(2) $(B, m, 1_B, \Delta, \varepsilon)$ 和 $(B, m, 1_B, \Delta^{\mathrm{cop}}, \varepsilon^{\mathrm{cop}})$ 诱导 $B \otimes B$ 上的弱 B-余环当且仅当 B 是弱双代数. 进一步, 映射 $\beta : B \otimes_{B_t} B \longrightarrow B \otimes B$ 定义为 $\beta(a \otimes b) = (a \otimes 1)\Delta(b)$ 是双射当且仅当 B 是弱 Hopf 代数.

证明 (1) 由例 3.1.5(1)(ii) 和 (iii), 充分性显然. 关于必要性, 由 (3.1.5), 容易看出 Δ 是代数同态; 由 (3.1.7), 可得 A 是余结合余代数; 对于 α, 由 (3.1.8), 有

$$1_1 \otimes 1_2 c = 1_1 \varepsilon(1_2 c_1) \otimes c_2,$$

一方面, 把上式应用到 $c = 1_1$ 上, 得

$$1_1 \otimes 1_2 1_1' \otimes 1_2' = 1_1 \varepsilon(1_2 1_1') \otimes 1_2' \otimes 1_3'.$$

另一方面, 把上式应用到 $c = 1$ 上, 得

$$1_1 \otimes 1_2 = 1_1 \varepsilon(1_2 1_1') \otimes 1_2',$$

那么, 在该式中, 对第二个因子取余乘 Δ, 得

$$\Delta^2(1) = 1_1 \otimes 1_2 1_1' \otimes 1_2'.$$

类似地, 由 β 可证 $\Delta^2(1) = 1_1 \otimes 1_1' 1_2 \otimes 1_2'$.

由 (3.1.6), 由 α, β 可得 $\varepsilon(abc) = \varepsilon(ab_1)\varepsilon(b_2 c) = \varepsilon(ab_2)\varepsilon(b_1 c)$.

(2) 易证.

在弱 Hopf 代数中有映射 $\varepsilon_s, \varepsilon_t : H \longrightarrow H$,

$$\varepsilon_s(x) = 1_1 \varepsilon(x 1_2) = S(x_1) x_2, \quad \varepsilon_t(x) = \varepsilon(1_1 x) 1_2 = x_1 S(x_2),$$

ε_s 称为**源余单位**(source counit) 映射, ε_t 称为**靶余单位**(target counit) 映射, 它们的像

$$H_s = \varepsilon_s(H) = \{x \in H \mid \Delta(x) = 1_1 \otimes x1_2 = 1_1 \otimes 1_2 x\}.$$

$$H_t = \varepsilon_t(H) = \{x \in H \mid \Delta(x) = 1_1 x \otimes 1_2 = x1_1 \otimes 1_2\}$$

分别称为源空间和靶空间. 对于余单位映射来说有下面重要的性质.

命题 3.1.7 对于任意的 $h, g \in H$, 有

(i) 余单位映射在 $\mathrm{End}_k(H)$ 中是幂等的, 即 $\varepsilon_t(\varepsilon_t(h)) = \varepsilon_t(h)$, $\varepsilon_s(\varepsilon_s(h)) = \varepsilon_s(h)$.

(ii) $(\mathbf{id} \otimes \varepsilon_t)\Delta(h) = 1_1 h \otimes 1_2$, $(\varepsilon_s \otimes \mathbf{id})\Delta(h) = 1_1 \otimes h1_2$.

(iii) 若 $\Delta(h) = 1_1 h \otimes 1_2 (\Delta(h) = 1_1 \otimes h1_2)$, 则有 $h = \varepsilon_t(h)(h = \varepsilon_s(h))$.

(iv) $\varepsilon_t \circ \varepsilon_s = \varepsilon_s \circ \varepsilon_t$.

(v) $h\varepsilon_t(g) = \varepsilon(h_1 g)h_2$, $\varepsilon_s(h)g = h_1\varepsilon(gh_2)$.

证明 这里我们只证 ε_t 这种情形. 由弱 Hopf 代数的定义可得

$$\varepsilon_t(\varepsilon_t(h)) = \varepsilon(1_1 h)\varepsilon(1_1' 1_2)1_2' = \varepsilon(1_1 h)\varepsilon(1_2)1_3 = \varepsilon_t(h),$$

这里 $1' = 1$ 均表示单位元. 故 (i) 成立.

$$h_1 \otimes \varepsilon_t(h_2) = h_1\varepsilon(1_1 h_2) \otimes 1_2 = 1_1 h_1 \varepsilon(1_2 h_2) \otimes 1_3 = 1_1 h \otimes 1_2.$$

故 (ii) 成立. 因为

$$\Delta(\varepsilon_t(h)) = \varepsilon(1_1 h)1_2 \otimes 1_3 = \varepsilon(1_1 h)1_1' 1_2 \otimes 1_2' = 1_1' \varepsilon_t(h) \otimes 1_2',$$

故有等式 $\Delta(h) = 1_1 h \otimes 1_2$, 两边作用 $(\varepsilon \otimes \mathbf{id})$, 可得 $h = \varepsilon_t(h)$. 即 (iii) 成立. 由等式 $1_1 \otimes 1_1' 1_2 \otimes 1_2' = 1_1 \otimes 1_2 1_1' \otimes 1_2'$ 直接可得 (iv). 因为

$$\varepsilon(h_1 g)h_2 = \varepsilon(h_1 g_1)h_2 \varepsilon_t(g_2) = \varepsilon(h_1 g_1)h_2 g_2 S(g_3)$$
$$= hg_1 S(g_2) = h\varepsilon_t(g).$$

故 (v) 成立.

命题 3.1.8 $H_t(H_s)$ 是 H 的左 (右) 余理想子代数, $H_t \circ H_s = H_s \circ H_t$. 并且

$$H_t = \{(\phi \otimes \mathbf{id})\Delta(1) \mid \phi \in H^*\}, \qquad H_s = \{(\mathbf{id} \otimes \phi)\Delta(1) \mid \phi \in H^*\}.$$

证明 由命题 3.1.7(iii) 和 (iv), 知 H_t 和 H_s 是余理想, 且是可交换的, 我们有

$$H_t = \varepsilon(1_1 H)1_2 \subset \{(\phi \otimes \mathbf{id})\Delta(1) \mid \phi \in H^*\},$$

反过来, $\phi(1_1)1_2 = \phi(1_1)\varepsilon_t(1_2) \subset H_t$, 因此 $H_t = \{(\phi \otimes \mathbf{id})\Delta(1) \mid \phi \in H^*\}$.

3.1 弱 Hopf 代数定义

下面说明 H_t 是代数: 对于任意的 $h,g \in H$, 有 $1 = \varepsilon_t(1) \in H_t$,

$$\varepsilon_t(h)\varepsilon_t(g) = \varepsilon(\varepsilon_t(h)_1 g)\varepsilon_t(h)_2$$
$$= \varepsilon(1_1\varepsilon_t(h)g)1_2 = \varepsilon_t(\varepsilon_t(h)g) \in H_t.$$

同理可证 H_s 也是代数.

若 H 是一个有限维弱 Hopf 代数, 则其对偶空间 $H^* = \mathrm{Hom}(H,k)$ 也是弱 Hopf 代数, 其上的结构映射为

$$\langle \phi\varphi, x \rangle = \langle \phi \otimes \varphi, \Delta(x) \rangle,$$
$$\langle \Delta(\phi), x \otimes y \rangle = \langle \phi, xy \rangle,$$
$$\langle S(\phi), x \rangle = \langle \phi, S(x) \rangle.$$

其中 $\phi, \varphi \in H^*, x, y \in H$. H^* 的单位为 ε, 余单位为 $\phi \mapsto \langle \phi, 1 \rangle$. 这里我们仍然使用 Sweedler 记法, 记

$$x \rightharpoonup \phi = \phi_1 \langle \phi_2, h \rangle, \quad \phi \leftharpoonup x = \langle \phi_1, x \rangle \phi_2.$$

例 3.1.9 (i) 若 $(H, m, u, \Delta, \varepsilon, S)$ 为有限维弱 Hopf 代数, 则 $H^{\mathrm{op}} = (H, m^{\mathrm{op}}, u, \Delta, \varepsilon, S^{-1})$, $H^{\mathrm{cop}} = (H, m, u, \Delta^{\mathrm{cop}}, \varepsilon, S^{-1})$, $H^{\mathrm{opcop}} = (H, m^{\mathrm{op}}, u, \Delta^{\mathrm{cop}}, \varepsilon, S)$ 也是弱 Hopf 代数.

(ii) 域 k 上全矩阵代数 M_n 为余交换有限维弱 Hopf 代数, 若对于任意矩阵单位 $e_{ij} \in M_n$ 满足:

$$\Delta(e_{ij}) = e_{ij} \otimes e_{ij}, \quad \varepsilon(e_{ij}) = 1, \quad S(e_{ij}) = e_{ji}.$$

G 有限, 则对偶的弱 Hopf 代数 $(kG)^*$ 与 G 上的代数函数同构, 也就是说其由幂等元 $P_g, g \in G$ 生成, 满足 $P_g P_h = \delta_{g,h} P_g$.

(iii) (**群胚代数**) G 为有限群胚, 则群胚代数 kG 是弱 Hopf 代数, 其中

$$\Delta(g) = g \otimes g, \quad \varepsilon(g) = 1, \quad S(g) = g^{-1}, \quad \forall g \in G.$$

此时, $(kG)^*$ 也为弱 Hopf 代数, 其中

$$\Delta(P_g) = \sum_{uv=g} P_u \otimes P_v, \quad \varepsilon(P_g) = \delta_{g,gg^{-1}}, \quad S(P_g) = P_{g^{-1}}.$$

(iv) 设 H 是一个 Hopf 代数, B 关于作用 $b \otimes h \mapsto b \cdot h$ 为**可分的**(separable) 右 H-模代数. 则 B^{op} 关于模作用 $h \cdot a = a \cdot S(h)$ 为左模代数, 则向量空间 $B^{\mathrm{op}} \otimes H \otimes B$ 关于乘法

$$(a \otimes h \otimes b)(a' \otimes h' \otimes b') = (h_1 \cdot a')a \otimes h_2 h'_1 \otimes (b \cdot h'_2)b', \quad \forall a, a' \in B^{\mathrm{op}}, b, b' \in B, h, h' \in H$$

构成一个代数，称为双边交叉积代数 (double crossed product algebra)，记为 $B^{\mathrm{op}}\#H\#B$. 令 e 为 B 中的可分元, $w \in B^*$ 由 $(w \otimes \mathbf{id})e = (\mathbf{id} \otimes w)e = 1$ 唯一决定. 易证 w 是 B 的**左正规表示**(left regular representation) 的**迹**(trace), 并且满足

$$w((h\cdot a)b) = w(a(b\cdot h)), \quad e^{(1)} \otimes (h\cdot e^{(2)}) = (e^{(1)}\cdot h) \otimes e^{(2)},$$

其中 $a \in B^{\mathrm{op}}, e = e^{(1)} \otimes e^{(2)}$.

则 $B^{\mathrm{op}}\#H\#B$ 关于结构

$$\Delta(a \otimes h \otimes b) = (a \otimes h_1 \otimes e^{(1)}) \otimes ((h_2 \cdot e^{(2)}) \otimes h_3 \otimes b),$$
$$\varepsilon(a \otimes h \otimes b) = w(a(h\cdot b)) = w(a(b\cdot S(h))),$$
$$S(a \otimes b \otimes h) = b \otimes S(h) \otimes a$$

是一个弱 Hopf 代数.

例 3.1.10 (Temperley-Lieb 代数) 设 $k = \mathbb{C}$ 是复数域, $\lambda^{-1} = 4\cos^2\dfrac{\pi}{n+3}(n \geqslant 2)$, $e_1, e_2, \cdots$ 是一列幂等元, 对任意 i,j 满足辫子像 (braid-like) 关系:

$$e_i e_{i\pm 1} e_i = \lambda e_i,$$
$$e_i e_j = e_j e_i, \quad |x-y| \geqslant 2.$$

设 $A_{k,l}$ 是由 $1, e_k, e_{k+1}, \cdots, e_l (k \leqslant l)$ 生成的代数, σ 是 $H = A_{1,2n-1}$ 的代数反同态, $\sigma(e_i) = e_{2n-i}$. $P_k \in A_{2n-k,2n-1} \otimes A_{1,k}$ 是 $A_{1,k}$ 中的可分幂等元在映射 $(\sigma \otimes \mathbf{id})$ 下的像. 记 τ 为 H 上的非退化 Markov 迹, w 表示 τ 在 $A_{n+1,2n-l}$ 上的限制指标, 即 $A_{n+1,2n-l}$ 中唯一的中心元满足 $\tau(w)$ 等于 $A_{n+1,2n-l}$ 左正规表示的迹.

H 是一个弱 Hopf 代数, 其上的结构为

$$\Delta(yz) = (z \otimes y)P_{n-1}, \quad y \in A_{n+1,2n-l}, \ z \in A_{1,n-l},$$
$$\Delta(e_n) = (1 \otimes w)P_n(1 \otimes w^{-1}),$$
$$S(h) = w^{-1}\sigma(h)w,$$
$$\varepsilon(h) = \lambda^{-n}\tau(hfw), \quad h \in A.$$

其中 $f = \lambda^{\frac{n(n-1)}{2}}(e_n e_{n-1} \cdots e_1)(e_{n+1} e_n \cdots e_2) \cdots (e_{2n-1} e_{2n-2} \cdots e_n)$ 为 Jones 投射, 对应 n-步基础构造 (n-step basic construction).

3.2 积分性质

定义 3.2.1 H 为弱 Hopf 代数, $l \in H(r \in H)$ 称为 H 的**左 (右) 积分** (left (right) integral), 若对任意的 $x \in H$ 满足:

$$xl = \varepsilon_t(x)l \quad (rx = r\varepsilon_s(x)).$$

3.2 积分性质

我们用 $\int_H^l \left(\int_H^r\right)$ 表示左 (右) 积分空间, $\int_H := \int_H^l \cap \int_H^r$ 为**双边积分空间**. H 的左 (右) 积分称为**非退化的**若其在 H 的对偶空间 H^* 上定义了一个非退化函数. H 的左 (右) 积分称为**正规的**, 若 $\varepsilon_t(x) = 1 (\varepsilon_s(x) = 1)$.

左 (右) 积分对偶的概念即为**左 (右) 不变测度**(invariant measure), 即 $\phi \in H^*$ 称为 H 的左 (右) 不变测度 (invariant measure), 若 $(\mathrm{id} \otimes \phi)\Delta = (\varepsilon_t \otimes \phi)\Delta ((\phi \otimes \mathrm{id})\Delta = (\phi \otimes \varepsilon_s)\Delta)$. 左 (右) 不变测度称为正规的, 若 $(\mathrm{id} \otimes \phi)\Delta = 1 ((\phi \otimes \mathrm{id})\Delta = 1)$.

例 3.2.2 (i) 设 G^0 是有限群胚的单位组成的集合, 则对于任意 $e \in G^0$, 有 $l_e = \sum_{gg^{-1}=e} g \left(r_e = \sum_{g^{-1}g=e} g\right)$ 张成 $\int_{kG}^l \left(\int_{kG}^r\right)$.

(ii) 若 $H = (kG)^*$, 则 $\int_H^l = \int_H^r = \mathrm{span}\{p_e \mid e \in G^0\}$.

命题 3.2.3 设 H 为弱 Hopf 代数. 则对任意 $x \in H$, 下列等价:

(i) $x \in \int_H^l$.

(ii) $x_1 \otimes yx_2 = S(y)x_1 \otimes x_2, \forall y \in H$.

(iii) $x \rightharpoonup H^* \subset H_t^*$.

(iv) $(\alpha \leftharpoonup y) \rightharpoonup x = S(y)(\alpha \rightharpoonup x), \forall \alpha \in H^*, y \in H$.

(v) $\mathrm{Ker}(\varepsilon_t)x = 0$.

(vi) $S(x) \in \int_H^r$.

对于弱 Hopf 代数 H 来说, 它既是代数也是余代数, 故我们可以考虑 H 上的模和余模, 类似于 Hopf 代数理论中的 Hopf 模定义, 下面我们给出弱 Hopf 代数 H 上的定义.

定义 3.2.4 H 为弱 Hopf 代数, $(M, \cdot, \rho)$ 称为 H 上的**右 H-Hopf 模**, 若 $(M, \cdot)$ 为右 H-模, (M, ρ) 为右 H-余模, 并且满足相容条件:

$$\rho(m \cdot x) = m_{(0)} \cdot x_1 \otimes m_{(1)}x_2$$

对于任意 $x \in H, m \in M$. 其中 $\rho(m) = m_{(0)} \otimes m_{(1)}$.

例 3.2.5 当 $\rho = \Delta$ 时, H 为 H-Hopf 模.

例 3.2.6 H^* 关于结构

$$\phi \cdot h = S(h) \rightharpoonup \phi, \qquad \phi_{(0)}\langle \varphi, \phi_{(1)}\rangle = \varphi\phi$$

为右 H-Hopf 模. 其中 $\phi, \varphi \in H^*, h \in H$.

例 3.2.7 $\Delta(1)(H \otimes H)\Delta(1)$ 是一个右弱 Hopf 模，其中模作用和余模作用为

$$(x \otimes y) \cdot z = xz_1 \otimes yz_2, \quad x \otimes y \longmapsto x \otimes y_1 \otimes y_2.$$

类似于 Hopf 代数，下面给出弱 Hopf 代数的 Hopf 模基本定理.

定理 3.2.8 H 为弱 Hopf 代数，M 为右 H-Hopf 模，$N = M^{coH} = \{m \in M \mid \rho(m) = m \cdot 1_1 \otimes 1_2\}$ 为 M 的余不变量空间. 则 H_t-模张量积 $N \otimes_{H_t} H$ (N 为右 H_t 子模) 关于作用

$$(m \otimes h) \cdot g = m \otimes hg,$$
$$\rho(m \otimes h) = (m \otimes h_1) \otimes h_2$$

为右 H-Hopf 模，其中 $m \in N, h, g \in H$. 特别地，映射

$$N \otimes_{H_t} H \longrightarrow M, \quad m \otimes h \longmapsto m \cdot h$$

为右 H-Hopf 模同构.

证明 易证 α 是 Hopf 模同态. 构造映射：

$$\beta : M \longrightarrow N \otimes_{H_t} H : m \mapsto (m_{(0)} \cdot S(m_{(1)})) \otimes m_{(2)}.$$

下面说明 β 是 α 的逆映射.

首先有 $m_{(0)} \cdot S(m_{(1)}) \in N$，因为

$$\begin{aligned}\rho(m_{(0)} \cdot S(m_{(1)})) &= (m_{(0)} \cdot S(m_{(3)})) \otimes m_{(1)} S(m_{(2)}) \\ &= (m_{(0)} \cdot S(1_2 m_{(1)})) \otimes S(1_1) \\ &= (m_{(0)} \cdot S(m_{(1)})1_1) \otimes 1_2.\end{aligned}$$

所以 β 的映射像为 $N \otimes_{H_t} H$.

其次 β 既是模同态，也是余模同态：

$$\begin{aligned}\beta(m \cdot h) &= m_{(0)} \cdot h_1 S(m_{(1)} h_2) \otimes m_{(2)} h_3 \\ &= m_{(0)} \cdot \varepsilon_t(h_1) S(m_{(1)}) \otimes m_{(2)} h_2 \\ &= m_{(0)} \cdot S(m_{(1)} 1_1) \otimes m_{(2)} 1_2 h = \beta(m) \cdot h, \\ \beta(m_{(0)}) \otimes m_{(1)} &= m_{(0)} \cdot S(m_{(1)}) \otimes m_{(2)} \otimes m_{(3)} \\ &= \beta(m)_{(0)} \otimes \beta(m)_{(1)}.\end{aligned}$$

最后证明 $\alpha \circ \beta = \mathbf{id}$, $\beta \circ \alpha = \mathbf{id}$:

$$\alpha \circ \beta(m) = m_{(0)} \cdot S(m_{(1)}) m_{(2)} = m_{(0)} \cdot \varepsilon_s(m_{(1)})$$
$$= m_{(0)} \cdot 1_1 \varepsilon(m_{(1)} 1_2) = m_{(0)} \varepsilon(m_{(1)}) = m,$$
$$\beta \circ \alpha(n \otimes h) = \beta(n \cdot h) = \beta(n) \cdot h$$
$$= n \cdot 1_1 S(1_2) \otimes 1_3 h = n \otimes h.$$

注记 3.2.9 $\int_{H^*}^l = H^{*\mathrm{co}H}$, 其中 H^* 作为 H-Hopf 模. 则条件 $\lambda_1 \otimes \lambda_2 = \lambda \cdot 1_1 \otimes 1_2$ 等价于 $\phi\lambda = (S(1_1) \rightharpoonup \lambda)\langle \phi, 1_2 \rangle = \varepsilon_t(\phi)\lambda$.

推论 3.2.10 $H^* \cong \int_{H^*}^l \otimes_{H_t} H$ 是右 H-Hopf 模同构. 特别地, $\int_{H^*}^l$ 是 H^* 的非零子空间.

证明 定理 3.2.8 中取 $M = H^*$ 及利用注记 3.2.9 可得.

由于弱 Hopf 代数中积分的存在性, 我们可以推广 Hopf 代数理论中非常著名的 Maschke 定理.

定理 3.2.11 设 H 为弱 Hopf 代数. 则下列条件等价:

(i) H 是半单的.

(ii) H 存在一个正规的左积分.

(iii) H 是可裂的 (separable).

证明 (i)$\Rightarrow$(ii): 假设 H 是半单的, 因为 $\mathrm{Ker}(\varepsilon_t)$ 是 H 的左理想, 则对于幂等元 p 有 $\mathrm{Ker}(\varepsilon_t) = Hp$, 所以由命题 3.2.3(iv) 可得 $\mathrm{Ker}\varepsilon_t(1-p)$ 和 $l = 1 - p$ 是左积分. 又因为 $\varepsilon_t(l) = 1 - \varepsilon_t(p) = 1$, 故左积分是正规的.

(ii) $\Rightarrow$(iii): 假设 l 是正规的, 由命题 3.2.3(ii) 可得 $l_1 \otimes S(l_2)$ 是 H 中的可裂元.

(iii) $\Rightarrow$(i): 直接可证.

设 H 是有限维弱 Hopf 代数, $l \in \int_H^l$. 如果存在 $\lambda \in H^*$, 使得 $\sum \lambda(l_2)l_1 = 1$, 则 $\lambda \in \int_{H^*}^l$ 是唯一的, 并且 $\sum \lambda_2(l)\lambda_1 = 1_{H^*}$. 称 (l, λ) 为**对偶左积分配对**. 类似地, 可定义**对偶右积分配对**.

3.3 Drinfel'd (余) 偶

设 H 为有限维弱 Hopf 代数, 下面定义 H 的 Drinfel'd 偶, 记为 $D(H)$. 考虑向量空间 $H^{*\mathrm{op}} \otimes H$, 乘法定义为

$$(\phi \otimes x)(\varphi \otimes y) = \varphi_2 \phi \otimes x_2 y \langle S(x_1), \varphi_1 \rangle \langle x_3, \varphi_3 \rangle,$$

其中 $\phi, \varphi \in H^{*\mathrm{op}}, x, y \in H$ 易知由元素 $\phi \otimes yx - (\varepsilon \leftharpoonup y)\phi \otimes x, y \in H_t$ 和 $\phi \otimes zx - (z \rightharpoonup \varepsilon)\phi \otimes x, z \in H_s$ 线性扩张的空间 J 为 $H^{*\mathrm{op}} \otimes H$ 的双边理想. 令 $D(H)$ 为**因子代数**(factor-algebra)$(H^{*\mathrm{op}} \otimes H)/J$. 令 $[\phi \otimes x]$ 表示 $\phi \otimes x$ 在 $D(H)$ 中所代表的类.

下面给出 H 的 **Drinfel'd 偶** $D(H)$ 的定义, 即如下定理.

定理 3.3.1 $D(H)$ 为弱 Hopf 代数, 其中单位元为 $[\varepsilon \otimes 1]$, 余乘、余单位和对极分别为

$$\Delta([\phi \otimes x]) = [\phi_1 \otimes x_1] \otimes [\phi_2 \otimes x_2],$$
$$\varepsilon([\phi \otimes x]) = \langle \varepsilon_t(x), \phi \rangle,$$
$$S([\phi \otimes x]) = [S^{-1}(\phi_2) \otimes S(x_2)]\langle x_1, \phi_1 \rangle \langle S(x_3), \phi_3 \rangle.$$

证明 参考文献 [34].

当 H 为一般的 Hopf 代数时, 上述定理退化成 Hopf 代数的 Drinfel'd 偶.

定义 3.3.2 拟三角弱 Hopf 代数是一个二元对 (H, R), 其中 H 是弱 Hopf 代数, $R = R^{(1)} \otimes R^{(2)} \in \Delta^{\mathrm{cop}}(1)(H \otimes H)\Delta(1)$ 满足:

(i) $\Delta^{\mathrm{cop}}(x)R = R\Delta(x)$.

(ii) $(\mathbf{id} \otimes \Delta)(R) = R_{13}R_{12}, (\Delta \otimes \mathbf{id})(R) = R_{13}R_{23}$, 其中 $R_{12} = R \otimes 1, R_{23} = 1 \otimes R$.

(iii) 存在 $\overline{R} \in \Delta(1)(H \otimes H)\Delta^{\mathrm{cop}}(1)$, 使得 $R\overline{R} = \Delta^{\mathrm{cop}}(1), \overline{R}R = \Delta(1)$.

注记 3.3.3 $\overline{R}$ 是由 R 唯一决定的: 若存在 $\overline{R'} \in \Delta(1)(H \otimes H)\Delta^{\mathrm{cop}}(1)$, 则

$$\overline{R} = \overline{R}\Delta^{\mathrm{cop}}(1) = \overline{R}R\overline{R'} = \Delta(1)\overline{R'} = \overline{R'}.$$

引理 3.3.4 (H, R) 为拟三角弱 Hopf 代数, 则 R 满足量子 Yang-Baxter 方程:

$$R_{12}R_{13}R_{23} = R_{23}R_{13}R_{12}.$$

命题 3.3.5 对于任意的拟三角弱 Hopf 代数 (H, R), 有

$$(\varepsilon_s \otimes \mathbf{id})(R) = \Delta(1), \qquad (\mathbf{id} \otimes \varepsilon_s)(R) = (S \otimes \mathbf{id})\Delta^{\mathrm{cop}}(1),$$
$$(\varepsilon_t \otimes \mathbf{id})(R) = \Delta^{\mathrm{cop}}(1), \qquad (\mathbf{id} \otimes \varepsilon_t)(R) = (S \otimes \mathbf{id})\Delta(1),$$
$$(S \otimes \mathbf{id})(R) = (\mathbf{id} \otimes S^{-1})(R) = \overline{R}, \qquad (S \otimes S)(R) = R.$$

命题 3.3.6 设 (H, R) 为拟三角弱 Hopf 代数, 则对任意的 $h \in H$,

$$S^2(h) = uhu^{-1},$$

其中 $u = S(R^{(2)})R^{(1)}$ 是 H 的可逆元, 满足

$$u^{-1} = R^{(2)}S^2(R^{(1)}), \quad \Delta(u) = \overline{R}\,\overline{R}_{21}(u \otimes u).$$

3.3 Drinfel'd (余) 偶

u 称为 H 的 **Drinfel'd 元**.

命题 3.3.7 (H, R) 为拟三角弱 Hopf 代数, 考虑线性映射 $F: H^* \longrightarrow H$

$$F(\phi) = (\phi \otimes \mathbf{id})(R_{21}R), \qquad \forall \phi \in H^*.$$

则 $F(\phi) \in C(H_s)$, 其中 $C(H_s) = \{x \in H \mid xy = yx, \ \forall y \in H_s\}$.

定义 3.3.8 我们称拟三角弱 Hopf 代数为**可分解的**(factorizable), 若上述命题中的映射 $F: H^* \longrightarrow C(H_s)$ 为满射.

命题 3.3.9 Drinfel'd 偶 $D(H)$ 是拟三角弱 Hopf 代数, 其拟三角结构为

$$R = \sum_i [\xi^i \otimes 1] \otimes [\varepsilon \otimes f_i], \quad \overline{R} = \sum_j [S^{-1}(\xi^j) \otimes 1] \otimes [\varepsilon \otimes f_j].$$

其中 $\{f_i\} \in H, \{\xi^i\} \in H^*$ 是一组对偶基, 满足 $\langle \xi^i, f_j \rangle = \delta_{i,j}$.

证明 我们把 H^{*op} 和 H 分别与 $[H^{*op} \otimes 1], [\varepsilon \otimes H]$ 等同. 等式 $(\mathbf{id} \otimes \Delta)(R) = R_{13}R_{12}$, $(\Delta \otimes \mathbf{id})(R) = R_{13}R_{23}$ 可以写成

$$\sum_i \xi_1^i \otimes \xi_2^i \otimes f_i = \sum_{ij} \xi^i \otimes \xi^j \otimes f_i f_j,$$

$$\sum_i \xi^i \otimes f_{i1} \otimes f_{i2} = \sum_{ij} \xi^j \xi^i \otimes f_j \otimes f_i.$$

上述等式可以通过对等式两边的第二个元素赋值 $\phi \in H^{*op}$, 第三个元素赋值 $h \in H$ 得到.

下面证明 $R \circ \Delta = \Delta^{\text{cop}} \circ R$. 利用等式 $\sum_i \langle a, \xi^i \rangle f_i = a$ 和 $\sum_i \xi^i \langle f_i, \phi \rangle = \phi$, 有

$$R\Delta([\phi \otimes h]) = \sum_i [\phi_1 \xi^i \otimes h_1] \otimes [\phi_3 \otimes f_{i2}h_2] \langle S(f_{i1}), \phi_2 \rangle \langle f_{i3}, \phi_4 \rangle$$

$$= \sum_i [\phi_1 S(\phi_2) \xi^i \phi_4 \otimes h_1] \otimes [\phi_3 \otimes f_i h_2]$$

$$= \sum_i [\xi^i \phi_3 \otimes \langle 1_1, \varepsilon_t(\phi_1) \rangle 1_2 h_1] \otimes [\phi_2 \otimes f_i h_2]$$

$$= \sum_i [\xi^i \phi_2 \otimes \langle 1_1, \varepsilon_1 \rangle 1_2 h_1] \otimes [\varepsilon_2 \phi_1 \otimes f_i h_2]$$

$$= \sum_i [\xi^i \phi_2 \otimes h_2] \otimes [\langle \varepsilon_t(h_1), \varepsilon_1 \rangle \varepsilon_2 \phi_1 \otimes f_i h_3]$$

$$= \sum_i [\xi^i \phi_2 \otimes h_3] \otimes [\phi_1 \otimes h_1 S(h_2) f_i h_4]$$

$$= \sum_i [\xi_2^i \phi_2 \otimes h_3] \otimes [\phi_1 \otimes h_1 f_i] \langle S(h_2), \xi_1^i \rangle \langle h_4, \xi_3^i \rangle$$

$$= \Delta^{\text{cop}}([\phi \otimes h])R.$$

最后我们验证 $\overline{R} = \sum_j [S^{-1}(\xi^j) \otimes 1] \otimes [\varepsilon \otimes f_j]$ 满足 $\overline{R}R = \Delta(1), R\overline{R} = \Delta^{\mathrm{cop}}(1)$. $\overline{R}R = \Delta(1)$ 等价于

$$\sum_{i,j}[\xi^i S^{-1}(\xi^j) \otimes 1] \otimes [\varepsilon \otimes f_j f_i] = [\langle 1_1, \varepsilon_2' \rangle \varepsilon_1' \varepsilon_1 \otimes 1] \otimes [\varepsilon \otimes \langle 1_1', \varepsilon_2 \rangle 1_2' 1_2],$$

上述等式等价于 $H^{*\mathrm{op}} \otimes H$ 中的等式

$$\sum_{i,j} \xi^i S^{-1}(\xi^j) \otimes f_j f_i = \langle 1_1, \varepsilon_2' \rangle \varepsilon_1' \varepsilon_1 \otimes \langle 1_1', \varepsilon_2 \rangle 1_2' 1_2.$$

对等式两边第二个元素赋值 $\phi \in H^*$, 则

$$\begin{aligned}\phi_2 S^{-1}(\phi_1) &= \langle 1_1, \varepsilon_2' \rangle \varepsilon_1' \varepsilon_1 \langle 1_1', \varepsilon_2 \rangle \langle 1_2' 1_2, \phi \rangle \\ &= \varepsilon_s^{\mathrm{op}}(\phi_2) \varepsilon_s^{\mathrm{op}}(\phi_1),\end{aligned}$$

其中 $\varepsilon_s^{\mathrm{op}}(\phi) = \phi_2 S^{-1}(\phi_1)$ 是 $H^{*\mathrm{op}}$ 的源余单位映射. 类似地, 可得 $R\overline{R} = \Delta^{\mathrm{cop}}(1)$.

下面我们给出 Drinfel'd 偶 $D(H)$ 的对偶情形, 其对偶 $\overline{D(H)}$ 也是弱 Hopf 代数.

对偶的弱 Hopf 代数 $\overline{D(H)}$ 由元素 $\sum_k h_k \otimes \phi_k \in H \otimes H^{*\mathrm{op}}$ 组成. 满足 $\sum_k (h_k \otimes \phi_k)|_J = 0$. $\overline{D(H)}$ 上的结构为

$$\left(\sum_k h_k \otimes \phi_k\right)\left(\sum_l g_l \otimes \varphi_l\right) = \sum_{k,l} h_k g_l \otimes \phi_k \varphi_l,$$

$$1_{D(H)} = 1_2 \otimes (\varepsilon \leftharpoonup 1_1),$$

$$\Delta\left(\sum_k h_k \otimes \phi_k\right) = \sum_{i,j,k}(h_{k2} \otimes \xi^i \phi_{k1} \xi^j) \otimes (S(f_i) h_{k1} f_j \otimes \phi_{k2}),$$

$$\varepsilon\left(\sum_k h_k \otimes \phi_k\right) = \sum_k \varepsilon(h_k) \widehat{\varepsilon}(\phi_k),$$

$$S\left(\sum_k h_k \otimes \phi_k\right) = \sum_{i,j,k} f_i S^{-1}(h_k) S(f_j) \otimes \xi^i S(\phi) \xi^j,$$

其中 $\sum_k h_k \otimes \phi_k, \sum_l g_l \otimes \varphi_l \in \overline{D(H)}$, 且 $\widehat{\varepsilon}$ 为 $\overline{D(H)}$ 的余单位.

命题 3.3.10 Drinfel'd 偶 $D(H)$ 是可分解的.

证明 对于任意的 $g \in H, \psi \in H^*$, 元素

$$Q_{g \otimes \psi} = \Sigma_{k,l} f_k g f_l \otimes \xi^k \psi S(\xi^l) \in H \otimes H^{*\mathrm{cop}}$$

属于 $\overline{D(H)}$. 因为

$$\begin{aligned}\langle Q_{g \otimes \psi}, \phi \otimes zh \rangle &= \langle zh_1, \phi_1 \rangle \langle S(h_3), \phi_3 \rangle \langle g, \phi_2 \rangle \langle h_2, \psi \rangle \\ &= \langle h_1, (\phi \leftharpoonup z)_1 \rangle \langle S(h_3), (\phi \leftharpoonup z)_3 \rangle \langle g, (\phi \leftharpoonup z)_2 \rangle \langle h_2, \psi \rangle \\ &= \langle Q_{g \otimes \psi}, (\varepsilon \leftharpoonup z) \phi \otimes h \rangle,\end{aligned}$$

其中 $h \in H, \phi \in H^*, z \in H_t$. 同样地有

$$\langle Q_{g\otimes\psi}, \phi \otimes yh\rangle = \langle Q_{g\otimes\psi}, (y \rightharpoonup \varepsilon)\phi \otimes h\rangle.$$

利用等式

$$\sum_j S(\xi_1^j)\langle g, \xi_2^j\rangle \otimes \xi_3^j \otimes f_i = \sum_{i,j} S(\xi^i) \otimes \xi^j \otimes f_i g f_j,$$

$$\sum_{i,j} S(\xi^i)\xi^j \otimes f_i f_j = \varepsilon_1 \otimes 1_1 \langle 1_2, \varepsilon_2\rangle,$$

可得

$$\begin{aligned}(Q_{g\otimes\psi} \otimes \mathbf{id})(R_{21}R) &= \sum_{i,j} Q_{g\otimes\psi}([\xi_2^j \otimes f_{i2}])[\xi^i \otimes f_j]\langle S(f_{i1}), \xi_1^j\rangle\langle f_{i3}, \xi_3^j\rangle \\ &= \sum_{i,j} Q_{g\otimes\psi}([\xi_2^j \otimes f_i])[S(\xi_1^j)\xi^i\xi_3^j \otimes f_j] \\ &= \sum_{i,j,k,l} [S(\xi^j)\xi^k\psi S(\xi^l)\xi^i \otimes f_j f_k g f_l f_i] \\ &= [\varepsilon_1'\langle 1_2', \varepsilon_2'\rangle\psi\varepsilon_1\langle 1_2, \varepsilon_2\rangle \otimes 1_1' g 1_1] \\ &= [\psi\varepsilon_1\langle 1_2, \varepsilon_2\rangle \otimes g 1_1].\end{aligned}$$

因此

$$\begin{aligned}\langle 1_2, \varepsilon_2\rangle Q_{g1_1\otimes\psi\varepsilon_1}(R_{21}R) &= [\psi\varepsilon_1\varepsilon_1' \otimes g 1_1' 1_1]\langle 1_2, \varepsilon_2'\rangle\langle 1_2', \varepsilon_2\rangle \\ &= [\varepsilon_1 \otimes 1_1][\psi \otimes g]S([\varepsilon_2 \otimes 1_2]) = Ad_1^l([\psi \otimes g]).\end{aligned}$$

其中 $Ad_1^l(h) = 1_1 h S(1_2), \forall h \in H$. 可知

$$\overline{D(H)} \ni x \mapsto (Q_x \otimes \mathbf{id})(R_{21}R) \in C_{D(H)}(D(H)_s)$$

是一个满射, 故 $D(H)$ 是可分的.

3.4 深度 2 扩张

本节中关于扩张概念的参考文献见 1.5 节相关扩张的出处.

定义 3.4.1 有限维 C^* 代数称为**有限维弱** C^* **代数**, 若线性映射

$$\Delta: \mathcal{A} \longrightarrow \mathcal{A} \otimes \mathcal{A}, \quad \varepsilon: \mathcal{A} \longrightarrow \mathbb{C}, \quad S: \mathcal{A} \longrightarrow \mathcal{A}$$

对任意的 $x, y, z \in \mathcal{A}$ 满足下列条件:

(i) $\Delta(xy) = \Delta(x)\Delta(y)$.
(ii) $\Delta(x^*) = \Delta(x)^*$.
(iii) $(\Delta \otimes \mathbf{id}) \circ \Delta = (\mathbf{id} \otimes \Delta) \circ \Delta$.
(iv) $(\Delta(1) \otimes 1)(1 \otimes \Delta(1)) = (1 \otimes \Delta(1))(\Delta(1) \otimes 1) = 1_1 \otimes 1_2 \otimes 1_3$.
(v) $(\varepsilon \otimes \mathbf{id}) \circ \Delta = \mathbf{id} = (\mathbf{id} \otimes \varepsilon) \circ \Delta$.
(vi) $\varepsilon(xyz) = \varepsilon(xy_1)\varepsilon(y_2 z) = \varepsilon(xy_2)\varepsilon(y_1 z)$.
(vii) $S(x_1)x_2 = 1_1 \varepsilon(x 1_2)$.
(viii) $x_1 S(x_2) = \varepsilon(1_1 x) 1_2$.
(ix) $S(x_1)x_2 S(x_3) = S(x)$.

定义 3.4.2 一个 $*$-代数 $\mathcal{M}$ 称为 $\mathcal{A}$-模代数, 若其上有一个左作用 $\triangleright : \mathcal{A} \otimes \mathcal{M} \longrightarrow \mathcal{M}, (a \otimes m) \mapsto a \triangleright m$, 对于任意的 $a, b \in \mathcal{A}, m, n \in \mathcal{M}$ 满足:

$$(ab) \triangleright m = a \triangleright (b \triangleright m), \quad 1_{\mathcal{A}} \triangleright m = m,$$
$$a \triangleright (mn) = (a_1 \triangleright m)(a_2 \triangleright n), \quad (a \triangleright m)^* = a^* \triangleright m^*,$$
$$a \triangleright 1_{\mathcal{M}} = (a_1 S(a_2)) \triangleright 1_{\mathcal{M}} \equiv \varepsilon(1_1 a) 1_2 \triangleright 1_{\mathcal{M}}.$$

定义 3.4.3 $\mathcal{M}$ 是 $\mathcal{A}$-模代数, **固定点代数**(fixed point algebras) $\mathcal{N} \equiv \mathcal{M}^{\mathcal{A}} \subset \mathcal{M}$ 是一个有单位的 $*$-子代数, 并且对于 $n \in \mathcal{M}$, 满足下列其中之一:

(i) $a \triangleright (mn) = (a \triangleright m)n$.
(ii) $a \triangleright n = a_1 S(a_2) \triangleright n$.
(iii) $a \triangleright (nm) = n(a \triangleright m)$.
(iv) $a \triangleright n = a_2 S^{-1}(a_1) \triangleright n$.

对任意的 $a \in \mathcal{A}, m \in \mathcal{M}$.

定义 3.4.4 $\mathcal{M}$ 是 $\mathcal{A}$-模 $*$-代数, 下面定义**交叉积**(crossed product)$\mathcal{M} \# \mathcal{A}$: 作为向量空间有 $\mathcal{M} \# \mathcal{A} = \mathcal{M} \otimes_{\mathcal{A}_L} \mathcal{A}$, 其中 $\mathcal{A}_L$ 作用在 $\mathcal{A}$ 上通过左乘, 作用在 $\mathcal{M}$ 上通过右乘 (利用 $\mu_{\triangleright}$ 的像). $\mathcal{M} \# \mathcal{A}$ 上的 $*$-代数结构为

$$(m \# a)(m' \# a') = (m(a_1 \triangleright m') \# a_2 a'),$$

$$(m \# a)^* = (1_{\mathcal{M}} \# a^*)(m^* \# 1_{\mathcal{A}}) \equiv (a_1^* \triangleright m^* \# a_2^*).$$

其中 $m, m' \in \mathcal{M}\ a, a' \in \mathcal{A}$.

下面给出本节的主要结果.

引理 3.4.5 关于态 (state) $w \in \mathcal{M}$, 下列等价:

(i) w 是 $\mathcal{A}$-不变量, 即 $w = w \circ E_h$.
(ii) $w(a \triangleright m) = w((S^{-1}(a) \triangleright 1_{\mathcal{M}})m)$, $\forall a \in \mathcal{A}, m \in \mathcal{M}$.
(iii) $w(a \triangleright m) = w(m(S(a) \triangleright 1_{\mathcal{M}}))$, $\forall a \in \mathcal{A}, m \in \mathcal{M}$.

3.4 深度 2 扩张

命题 3.4.6 (i) 设 $l_0 := hg_L^{-1} \in \mathcal{L}(\mathcal{A})$, 则 $V : \mathcal{H}_w \longrightarrow \mathcal{H}_{\text{cros}}, V|m\rangle_w := |ml_0\rangle_{\text{cros}}$ 为等距 (isometry) 映射, 并且对任意 $m \in \mathcal{M}, a \in \mathcal{A}$ 有

$$V^*|ma\rangle_{\text{cros}} = |m\mu \triangleright (ag_L)\rangle_w,$$

$$VV^*|ma\rangle_{\text{cros}} = |mag_L l_0\rangle_{\text{cros}},$$

$$\pi_w(m) = V^* \pi_{\text{cros}}(m) V.$$

(ii) $\pi_{\text{cros}}(\mathcal{M}\#\mathcal{A})$ 中 $V\mathcal{H}_w \subset \mathcal{H}_{\text{cros}}$ 是不变的 (invariant). 因此 $\mathcal{H}_w$ 中, π_w 可以扩张成 $\mathcal{M}\#\mathcal{A}$ 的表示 (依然记为 π_w), 其结构为 $\pi_w(ma) := V^*\pi_{\text{cros}}(ma)V$. 满足

$$\pi_w(ma)|m'\rangle_w = |m(a \triangleright m')\rangle_w.$$

(iii) $\mathcal{M}h\mathcal{M} \subset \mathcal{M}\#\mathcal{A}$ 是一个理想, 且与 $\text{Ker}\pi_w$ 正交, 即

$$(\mathcal{M}h\mathcal{M})(\text{Ker}\pi_w) = (\text{Ker}\pi_w)(\mathcal{M}h\mathcal{M}) = 0.$$

我们将 $\mathcal{M}^{\mathcal{A}} \subset \mathcal{M}$ 的 Jones 扩张 (extension) 和理想 $\mathcal{M}_1 \subset \mathcal{M}\#\mathcal{A}$ 等同. 在 $\mathcal{B}(\mathcal{H}_w)$ 中, 令

$$\mathcal{M}_{-1} := \pi_w(\mathcal{M}^{\mathcal{A}}),$$

$$\mathcal{M}_0 := \pi_w(\mathcal{M}),$$

$$\mathcal{M}_1 := \pi_w(\mathcal{M}\#\mathcal{A}).$$

设 $l \in \mathcal{L}(A)$ 是正的 (positive) 正规的非退化左积分, $e_l = d_R(l)^{\frac{1}{2}} h d_R(l)^{\frac{1}{2}}$ 是可结合的 "Jones 投射", $\lambda_l \in \mathcal{L}(A)$ 记为 l 的 p-对偶.

以下给出构造基本定理.

定理 3.4.7 w 是 $\mathcal{M}$ 上正规的忠实的 (faithful) $\mathcal{A}$-不变的态. 设 $u_i \in \mathcal{M}$ 是 $E_h := \mathcal{M}_0 \longrightarrow \mathcal{M}_{-1}$ 的 Pimsner Popa 基, 其中 $h \in \mathcal{A}$ 是正规的 Haar 积分. 则

(i) 扩张 $\mathcal{M}_{-1} \subset \mathcal{M}_0 \subset \mathcal{M}_1$ 是含有有限指数的 Jones 三元组, 并且为深度 2 扩张.

(ii) $P := \sum_i u_i h u_i^* \in \mathcal{M}\#\mathcal{A}$ 为中心投射 (central projection), 满足 $\text{Ker}\pi_w = (1-p)(\mathcal{M}\#\mathcal{A})$, 因此 $\mathcal{M}_1 \cong p(\mathcal{M}\#\mathcal{A})$.

(iii) 由 $p(\mathcal{M}\#\mathcal{A}) = \mathcal{M}h\mathcal{M}$ 和映射 $m \mapsto pm$ 可得单位扩张 (unital inclusion) $\mathcal{M} \longrightarrow \mathcal{M}h\mathcal{M}$.

(iv) 对于正的非退化的正规的左积分 $l \in \mathcal{L}(A)$, Jones 投射 $\mathbf{e}_l \in \mathcal{M}_1$ 有条件期望 (conditional expectation) $E_l : \mathcal{M}_0 \longrightarrow \mathcal{M}_{-1}$:

$$\mathbf{e}_l = \pi_w(e_l)$$

和 (非正规的) 条件期望 $E_l' : \mathcal{M}_1 \longrightarrow \mathcal{M}_0$:

$$E_l'(\pi_w(ma)) = \pi_w(m)\pi_w(\widehat{E}_{\lambda_l}(ap)).$$

E_l' 是 E_l 的对偶.

(v) $E_l \leqslant \tau_\triangleright(\text{Ind}l)$, 等号成立当且仅当 $p = 1$, 即 $\mathcal{M}_1 \cong \mathcal{M}_0\mathcal{A}$.

3.5 对　　偶

现在从 3.1 节中弱余环的对偶研究起. 对任意一个弱 A-余环, 从 C 到 A 的所有 A-模映射集合有环结构. 首先我们有如下经典同构:

$$C^* := \text{Hom}_{-A}(C, A) \simeq \text{Hom}_{-A}(CA, A),$$
$$(AC)^* := \text{Hom}_{-A}(AC, A) \simeq \text{Hom}_{-A}(ACA, A),$$
$$^*C := \text{Hom}_{A-}(C, A) \simeq \text{Hom}_{A-}(AC, A),$$
$$^*(CA) := \text{Hom}_{A-}(CA, A) \simeq \text{Hom}_{A-}(ACA, A),$$
$$^*C^* := \text{Hom}_{AA}(C, A) \simeq \text{Hom}_{AA}(ACA, A) \simeq {^*C} \cap C^*.$$

命题 3.5.1 具有上述记号, 则

(i) C^* 的环结构为: 对任意 $f, g \in C^*$

$$f *_r g : C \xrightarrow{\Delta} CA \otimes_A C \xrightarrow{g \otimes \text{id}} A \otimes_A C \simeq AC \xrightarrow{f} A,$$

即 $(f *_r g)(c) = \sum f(g(c_1)c_2)$, ε 是 C^* 中的中心幂等元, 且 $(AC)^* = \varepsilon *_r C^*$.

(ii) $^*C$ 的环结构为: 对任意 $f, g \in C^*$

$$f *_l g : C \xrightarrow{\Delta} C \otimes_A AC \xrightarrow{\text{id} \otimes f} C \otimes_A A \simeq CA \xrightarrow{g} A,$$

即 $(f *_l g)(c) = \sum g(c_1 f(c_2))$, ε 是 $^*C$ 中的中心幂等元, 且 $(CA)^* = \varepsilon *_l C^*$.

(iii) $^*C^*$ 的环结构为: 对任意 $f, g \in C^*$

$$f * g : C \xrightarrow{\Delta} C \otimes_A A \otimes_A C \xrightarrow{g \otimes \text{id} \otimes f} A,$$

即 $(f * g)(c) = \sum g(c_1)f(c_2)$, 单位元 ε;

(iv) 如果 C 是余结合 R-余环, 那么环结构也是结合的.

注记 3.5.2 如果 C 是 R-余环, 那么 C^*, $^*C$ 和 $^*C^*$ 有单位元 ε.

下面给出弱 Hopf 代数的模结构和余模结构. 这里 H 均表示弱 Hopf 代数.

3.5 对偶

定义 3.5.3 代数 M 称为**左 H-模代数**, 若 M 是左 H-模, 模结构为 $h \otimes m \longmapsto h \cdot m$, 且对任意的 $m, n \in M, h \in H$ 满足

$$h \cdot (mn) = (h_1 \cdot m) \otimes (h_2 \cdot n), \quad h \cdot 1 = \varepsilon_t(h) \cdot 1.$$

定义 3.5.4 设 M, N 为左 H-模, 线性映射 $f: M \longrightarrow N$ 称为**左 H-模同态**, 若对任意的 $m \in M, h \in H$ 满足

$$f(h \cdot m) = h \cdot f(m)$$

对任意的 $h \in H, m \in M$.

定义 3.5.5 $M^H = \{m \in M \mid h \cdot m = \varepsilon_t(h) \cdot m\}$ 称为 H 在 M 上的**不变量集合**.

定义 3.5.6 代数 M 称为**左 H-余模代数**, 若 M 为左 H-余模, 余模结构为 $\rho(m) = m_{(-1)} \otimes m_{(0)}$, 且对任意的 $m, n \in M$ 满足

$$\rho(mn) = \rho(m)\rho(n), \quad \rho(1) = (\varepsilon_s \otimes \mathbf{id})\rho(1).$$

定义 3.5.7 设 M, N 为左 H-余模, 线性映射 $f: M \longrightarrow N$ 称为**左 H-余模同态**, 若满足

$$f(m)_{(-1)} \otimes f(m)_{(0)} = m_{(-1)} \otimes f(m_{(0)})$$

对任意的 $m \in M$.

定义 3.5.8 $M^{\mathrm{co}H} = \{m \in M \mid m_{(-1)} \otimes m_{(0)} = \varepsilon_s(m_{(-1)}) \otimes m_{(0)}\}$ 称为 H 在 M 上的**余不变量集合**.

类似地可定义右模代数和右余模代数.

例 3.5.9 (i) 子代数 $H_t(H_s)$ 关于模作用

$$x \cdot z = \varepsilon_t(xz)(x \cdot z = \varepsilon_s(zS(x))), \quad \forall x \in H, z_t \in H_t, z_s \in H_s$$

为左 H-模代数, 这个模作用称为**平凡的** (trivial).

(ii) $C(H_s) = \{x \in H \mid xh = hx, \forall h \in H_s\}$ 关于**伴随作用**

$$x \cdot h = x_1 h S(x_2), \quad \forall x \in H, h \in C(H_s)$$

为左 H-模代数.

(iii) H^* 关于模作用

$$x \rightharpoonup \phi = \phi_1 \langle \phi_2, x \rangle, \quad \forall x \in H, \phi \in H^*$$

为左 H-模代数.

注记 3.5.10 如果 H 是有限维的, 那么左 H-余模代数关于作用

$$m \cdot \phi = \langle \phi, m_{(-1)} \rangle m_{(0)}, \quad \forall \phi \in H^*, m \in M$$

为右 H^*-模代数.

定义 3.5.11 H 为弱 Hopf 代数, 具有可逆对极 S. A 为左 H-模代数. $A \otimes H$ 关于单位元 $1 \otimes 1$ 和乘法

$$(a \otimes x)(b \otimes y) = a(x_1 \cdot b) \otimes x_2 y, \quad \forall a, b \in A, x, y \in H$$

为结合代数, 称为 A 和 H 的**冲积**, 记为 $A \# H$.

H 为弱 Hopf 代数, A 为左 H-模代数, 则冲积 $A \# H$ 为左 H^*-模代数, 其中模作用为

$$\phi \cdot (a \# h) = a \# (\phi \rightharpoonup h), \quad \forall \phi \in H^*, h \in H, a \in A.$$

当 H 为一般的有限维 Hopf 代数, 存在同构 $(A \# H) \# H^* \cong M_n(A)$. 其中 $n = \dim H$. 下面我们将这个结果推广到弱 Hopf 代数上 [282].

引理 3.5.12 H 为弱 Hopf 代数, A 为左 H-模代数, $A \# H$ 关于乘法为右 A-模. 则映射 $\alpha : (A \# H) \# H^* \longrightarrow \text{End}(A \# H)_A$

$$\alpha((a \# h) \# \phi)(b \# y) = (a \# x)(b \# (\phi \rightharpoonup y)) = a(x_1 \cdot b) \# x_2 (\phi \rightharpoonup y)$$

为代数同构. 其中 $a, b \in A, x, y \in H, \phi \in H^*$.

证明 首先, 我们验证 α 定义是合理的. 对于任意 $z \in H_t, \xi \in H_t^*$, 利用 $(\xi \rightharpoonup 1) \in H_s$, 可得

$$\begin{aligned}
\alpha((x \# zh) \# \phi)(y \# g) &= x(zh_1 \cdot y) \# h_2 (\phi \rightharpoonup g) \\
&= x(z \cdot 1)(h_1 \cdot y) \# h_2 (\phi \rightharpoonup g) \\
&= \alpha((x \cdot z) \# h) \# \phi)(y \# g), \\
\alpha((x \# h) \# \xi \phi)(y \# g) &= x(h_1 \cdot y) \# h_2 (\xi \rightharpoonup 1)(\phi \rightharpoonup g) \\
&= \alpha((x \# h(\xi \rightharpoonup 1)) \# \phi)(y \# g) \\
&= \alpha((x \# (S(\xi) \rightharpoonup h)) \# \phi)(y \# g) \\
&= \alpha((x \# (h \cdot \xi)) \# \phi)(y \# g).
\end{aligned}$$

其次, 利用等式 $\phi \rightharpoonup zy = z(\phi \rightharpoonup y)$, 我们验证 $\alpha((x \# h) \# \phi) \in \text{End}(A \# H)_A$.

$$\begin{aligned}
\alpha((x \# h) \# \phi)(y \# zg) &= x(h_1 \cdot y) \# h_2 z(\phi \rightharpoonup g) \\
&= x(h_1 S(z) \cdot y) \# h_2 (\phi \rightharpoonup g) \\
&= \alpha((x \# h) \# \phi)((y \cdot z) \# g).
\end{aligned}$$

3.5 对偶

α 与右作用 $w \in A$ 是可交换的, 因为

$$\begin{aligned}\alpha((x\#h)\#\phi)((y\#g)\cdot w) &= \alpha((x\#h)\#\phi)(y(g_1\cdot w)\#g_2)\\ &= (x\#h)(y(g_1\cdot w)\#(\phi \rightharpoonup g_2))\\ &= (x\#h)(y\#(\phi \rightharpoonup g))(w\#1)\\ &= (\alpha((x\#h)\#\phi)(y\#g))\cdot w.\end{aligned}$$

最后, 我们证明 α 为代数同态.

$$\begin{aligned}&\alpha(((x\#h)\#\phi)((x'\#h')\#\phi'))(y\#g)\\ &= \alpha((x\#h)(x'\#(\phi_1 \rightharpoonup h'))\#\phi_2\phi')(y\#g)\\ &= (x\#h)(x'\#(\phi_1 \rightharpoonup h'))(y\#(\phi_2\phi' \rightharpoonup g))\\ &= (x\#h)(\phi\cdot((x'\#h')(y\#(\phi' \rightharpoonup g))))\\ &= \alpha(((x\#h)\#\phi)((x'\#h')(y\#(\phi' \rightharpoonup g))))\\ &= \alpha(((x\#h)\#\phi)\circ \alpha((x'\#h')\#\phi'))(y\#g)\end{aligned}$$

对于任意的 $x, x', y \in A, h, h', g \in H, \phi, \phi' \in H^*$.

引理 3.5.13 定义映射 $\beta : \mathrm{End}(A\#H)_A \longrightarrow (A\#H)\#H^*$

$$\beta : T \mapsto \sum_i T(1\#f_{i2})(1\#S^{-1}(f_{i1}))\#\xi^i,$$

其中 $\{f_i\}, \{\xi^i\}$ 是一组对偶基. 则 β 与引理 3.5.11 中的 α 互逆.

证明 我们只需验证 $\beta \circ \alpha = \mathbf{id}_{(A\#H)\#H^*}$, $\alpha \circ \beta = \mathbf{id}_{\mathrm{End}(A\#H)_A}$ 即可.

对于任意的 $x \in A, h \in H, \phi \in H^*$, 有

$$\begin{aligned}\beta \circ \alpha((x\#h)\#\phi) &= \sum_i(x(h_1\cdot 1)\#h_2(\phi \rightharpoonup f_{i2})S^{-1}(f_{i1}))\#\xi^i\\ &= \sum_i(x\#h\langle\phi, f_{i3}\rangle f_{i2}S^{-1}(f_{i1}))\#\xi^i\\ &= \sum_i(x\#h\langle\phi, 1_2 f_i\rangle 1_1)\#\xi^i\\ &= (x\#h(\phi_1 \rightharpoonup 1))\#\phi_2\\ &= (x\#h)\#\varepsilon_t(\phi_1)\phi_2 = (x\#h)\#\phi.\end{aligned}$$

同样地, 对于任意的 $T \in \mathrm{End}(A\#H)_A$, 有

$$\begin{aligned}
\alpha \circ \beta(T)(y\#g) &= \sum_i \alpha(T(1\#f_{i2})(1\#S^{-1}(f_{i1}))\#\xi^i)(y\#g) \\
&= \sum_i T(1\#f_{i2})(1\#S^{-1}(f_{i1}))(y\#(\xi^i \rightharpoonup g)) \\
&= \sum_i T(1\#f_{i3})((S^{-1}(f_{i2}) \cdot y)\#S^{-1}(f_{i1})g_1)\langle \xi^i, g_2\rangle \\
&= T(1\#g_4)((S^{-1}(g_3) \cdot y)\#S^{-1}(g_2)g_1) \\
&= T(1\#g_3)((S^{-1}(g_2) \cdot y)(\varepsilon_s(g_1) \cdot 1)\#1) \\
&= T(1\#g_2)((S^{-1}(g_11_2) \cdot y)(1_1 \cdot 1)\#1) \\
&= T(1\#g_2)((S^{-1}(g_1) \cdot y)\#1) \\
&= T((g_2 S^{-1}(g_1) \cdot y)\#g_3) \\
&= T((1_1 \cdot y)\#1_2 g) = T(y\#g).
\end{aligned}$$

故引理得证.

定理 3.5.14 对于任意的 H-模代数 A, 存在代数同构:

$$(A\#H)\#H^* \cong \mathrm{End}(A\#H)_A.$$

证明 由引理 3.5.11 和引理 3.5.12 直接可得.

推论 3.5.15 $H\#H^* \cong \mathrm{End}(H)_{H_t}$. 特别地, $H\#H^*$ 为半单 (semisimple) 代数.

证明 已知 $H \cong H_t\#H$, 其中 H_t 是平凡的 H-模代数. 在定理 3.5.13 中令 $A = H_t$, 则 H 是投射生成的 H_t-模, 满足 $H\#H^* \cong \mathrm{End}(H)_{H_t}$. 因此 H_t 和 $H\#H^*$ 是 Morita 等价的. 由于 H_t 作为可分代数是半单的, 故 $H\#H^*$ 是半单的.

第4章 Hopf 型代数

本章主要介绍各类 Hopf 代数概念及例子[54, 101, 115, 117, 138, 210, 211, 232, 276, 282]. 这些概念是目前国际该领域研究的热点问题之一, 也有着其理论意义和应用背景.

4.1 (弱) 左 (右)Hopf 代数

我们知道, 一个双代数 B 成为 Hopf 代数的条件是恒等映射 I_d 在卷积代数 $\mathrm{Hom}_k(B,B)$ 中可逆, 即存在 $S \in \mathrm{Hom}_k(B,B)$ 使得 $S * I_d = \mu\varepsilon = I_d * S$, 这时的 S 被称为 Hopf 代数 B 的对极.

本节我们由上述条件在双代数上给出一个比一般 Hopf 代数弱一些的代数结构, 主要文献有 [82, 164].

定义 4.1.1 一个双代数 B 被称为**左 Hopf 代数**(left Hopf algebra)若其有一个**左对极**(left antipode), 即存在 $S \in \mathrm{Hom}_k(B,B)$ 使得 $S * I_d = \mu\varepsilon$.

类似地, 一个双代数 B 被称为**右 Hopf 代数**(right Hopf algebra)若其有一个**右对极**(right antipode), 即存在 $S \in \mathrm{Hom}_k(B,B)$ 使得 $I_d * S = \mu\varepsilon$.

下面我们给出一个是左 Hopf 代数的例子.

例 4.1.2 已知一个 k-余代数 C, 对于 $i \geqslant 0$, 令 $V_{2i} = C$, $V_{2i+1} = C^{\mathrm{cop}}$, 得到一个余代数直和, $V = \sum_{i=0}^{\infty} \oplus V_i$, 令 $T(V)$ 是向量空间 V 上的张量代数.

此时 $\Delta: V \longrightarrow V \otimes V$ 扩张为代数映射 $\Delta: T(V) \longrightarrow T(V) \otimes T(V)$, $\varepsilon: V \longrightarrow k$ 扩张成了代数映射 $\varepsilon: T(V) \longrightarrow k$, 使得 $T(V)$ 具有了双代数的结构, 映射 $S: V \longrightarrow V$ $(S(V_0, V_1, V_2, V_3, \cdots) \longrightarrow (0, V_0, V_1, V_2, \cdots))$ 是 V 的一个余代数反同态, 扩张为 $T(V)$ 的一个双代数反同态.

此时我们定义 $T(V)$ 的两个理想 $I = \mathrm{span}\{\sum S(x_1)x_2 - \varepsilon(x)1 \mid x \in V\}$, $J = \mathrm{span}\{\sum x_1 S(x_2) - \varepsilon(x)1 \mid x \in V\}$. 令 $L = I + J$, 则此时 $H(C) = T(V)/L$ 是定义在 C 上面的自由 Hopf 代数.

定义 $K = I + T(V)S(I)T(V)$, 则 $H_l(C) = T(V)/K$ 是一个左 Hopf 代数且不是 Hopf 代数.

定理 4.1.3 设 B 是一个左 Hopf 代数, 那么 B 是一个 Hopf 代数, 如果下面任何一个条件成立:

(1) B 是一个有限维的.

(2) B 是一个交换代数.

(3) B 是一个有点余代数.

(4) B 的余根 B_0 是余交换余代数.

命题 4.1.4 (1) 一个双代数可能有无限个左反对极;

(2) 左 Hopf 代数中的左反对极不一定是代数反同态.

注记 4.1.5 类似地, 我们不难给出弱左 (右)Hopf 代数的定义.

4.2 (弱) 双 Frobenius 代数

本节概念目前最具有概括性的文献见 [41, 198].

定义 4.2.1 令 A 是有限维代数, 记 $A^* = \mathrm{Hom}_k(A, k)$ 为 A 的对偶空间. 则 A^* 上自然有 A-双模结构, 其双模结构定义为 $(af)(b) = f(ba), (fa)(b) = f(ab)$, 任取 $f \in A^*, a, b \in A$. 如果有左 A-模同构 $_AA \cong (A_A)^*$ 或者等价于右 A-模同构 $_AA \cong (_AA)^*$, 则称 A 为**Frobenius 代数**(Frobenius algebra).

定义 4.2.2 令 C 是有限维余代数, 其余乘记为 Δ, 余单位记为 ε. 令 $C^* = \mathrm{Hom}_k(C, k)$ 为余代数 C 的对偶代数, 定义

$$fc = \sum c_1 f(c_2), \quad cf = \sum f(c_1)c_2$$

对任意 $f \in C^*, c \in C$, 则 C 成为 C^*-双模. 如果有 $C = tC^*$ 或 $C = C^*t$, 则称 (C, t) 为**Frobenius 余代数**(Frobenius coalgebra). 更多关于 Frobenius 余代数的相关知识请参阅文献.

定义 4.2.3 设 A 是有限维代数和余代数, 令 $\phi \in A^*$ 和 $t \in A$. 如果我们有 (A, ϕ) 是 Frobenius 代数, 同时 (A, t) 是 Frobenius 余代数, 则称三元素 (A, ϕ, t) 为**预双 Frobenius 代数**(prefrobenius algebra)[64], 如果 (A, ϕ, t) 是预双 Frobenius 代数, 则存在如下的线性映射, $\psi : A \longrightarrow A, \psi(a) = \sum \phi(t_1 a)t_2, a \in A$.

最后我们给出双 Frobenius 代数的定义.

定义 4.2.4 设 (A, ϕ, t) 是预双 Frobenius 代数, 如果 A 满足下列条件:

(i) 1_A 是群样元并且 ε 是代数映射;

(ii) $\psi : A \longrightarrow A$ 既是反代数映射又是反余代数映射,

则称四元素 (A, ϕ, t, ψ) 为**双 Frobenius 代数**(double frobenius algebra), 其中 ψ 称为双 Frobenius 代数 (A, ϕ, t, ψ) 的对极.

下面给出一类双 Frobenius 代数的例子.

例 4.2.5 令 $A = A_0 \oplus A_1 \oplus A_2 \oplus A_3 = k1_A \oplus (kx_1 \oplus kx_2 \oplus \cdots \oplus kx_n) \oplus (ky_1 \oplus ky_2 \oplus \cdots \oplus ky_n) \oplus kz$ 是 Z-分次的向量空间, 其中 n 取正整数. 令 $\lambda_{i,j} \in k$, 定义 $\Lambda = (\lambda_{i,j})_{n \times n}$ 是 $n \times n$ 矩阵, 则在 A 中可以定义如下的乘法: $x_i y_j = \delta_{i,j} z, y_i x_j = \lambda_{i,j} z, x_2^i = y_2^i = z_2 = x_i z = z x_i = y_i z = z y_i = 0$, 其中 $i = 1, 2, \cdots, n$. 这里称矩阵

4.2 (弱) 双 Frobenius 代数

$\Lambda = (\lambda_{i,j})_{n \times n}$ 为 A 的结构矩阵. 为叙述方便, 取 x_i 和 y_i 的下指标为 Z_n 中的元素. 令 π 是 $(1, 2, \cdots, n)$ 上的置换. 在 A 上定义余乘 Δ 和余单位 ε 如下:

$$\Delta(1) = 1 \otimes 1, \quad \Delta(x_i) = 1 \otimes x_i + x_i \otimes 1, \quad \Delta(y_i) = 1 \otimes y_i + y_i \otimes 1,$$

$$\Delta(z) = 1 \otimes z + z \otimes 1 + \sum_{i \in Z_n} x_i \otimes y_{\pi(i)} + \sum_{i \in Z_n} y_{\pi(i)} \otimes x_i,$$

$\varepsilon(1) = 1, \varepsilon(x_i) = \varepsilon(y_i) = \varepsilon(z) = 0$, 任取 $i \in Z_n$, 记 I_n 是 n 阶单位阵, I_d 为恒等映射. 则 A 是双 Frobenius 代数当且仅当 $\Lambda = I_n, \pi^2 = I_d$. 特别地, 当 $n > 2$ 时, A 是非 Hopf 代数的双 Frobenius 代数.

例 4.2.6 令 A 是一个数域 k 上的有限维 Hopf 代数. 余单位记为 ε_A, 对极记为 S, 选定左积分 $\lambda \in A, \lambda^* \in A^*$ 使得 $\langle \lambda^*, \lambda \rangle = 1$. 映射 $\Phi: A \longrightarrow A^*$, $\Phi(a) = \lambda^* \leftharpoonup S_{-1}(a)$, 任取 $a \in A$. 其逆映射 Φ^{-1} 定义为 $\Phi^{-1}(a^*) = a^* \rightharpoonup \lambda$, 任取 $a^* \in A^*$. 定义: $a \star b = \Phi^{-1}(\Phi(a)\Phi(b))$, 任取 $a, b \in A$. 那么 A^* 是一个结合代数, 其上有单位元 λ. 其上的乘法为: $a \star b = \sum \langle \lambda^*, S^{-1}(a_1)b \rangle a_2 = \sum \langle \lambda^*, S^{-1}(a)b_2 \rangle b_1$. 定义 $\alpha \in A^*: \lambda a = \langle \alpha, a \rangle \lambda$, 则此时 $(A, \cdot, \star, \varepsilon = \varepsilon_A, \omega = \lambda^*, \sigma = S, \tau = S^{-1}, s = \alpha^{-1} \rightharpoonup \lambda, t = 1)$ 是一个双 Fronbenius 代数.

类似于由 Hopf 代数得到弱 Hopf 代数, 我们将双 Frobenius 代数公理进行弱化, 则可以得到另一种类 Hopf 代数: 弱双 Frobenius 代数[41], 其具体定义如下.

定义 4.2.7 设 H 既是代数又是余代数, $\psi \in H^*, t \in H$, 定义线性映射 $S: H \longrightarrow H$ 如下:

$$S(h) = t \leftharpoonup (h \rightharpoonup \psi) = \sum \psi(t_1 h) t_2.$$

称 (H, ψ, T, H) 为**弱双 Frobenius 代数**(weak bi-Frobenius algebra)当下面的条件成立:

(i) $\Delta(h) = \Delta(h)\Delta(1) = \Delta(1)\Delta(h)$, 任取 $h \in H$;

(ii) $(\Delta \otimes i_d)\Delta(1) = (\Delta(1) \otimes 1)(1 \otimes \Delta(1)) = (1 \otimes \Delta(1))(\Delta(1) \otimes 1)$;

(iii) $\sum \varepsilon(g_1 h_1) g_2 h_2 = gh = \sum (g_1 h_1) \varepsilon(g_2 h_2)$, 任取 $g, h \in H$;

(iv) $\varepsilon(fgh) = \sum \varepsilon(fg_1)\varepsilon(fg_2) = \sum \varepsilon(fg_2)\varepsilon(g_1)h$, 任取 $f, g, h \in H$;

(v) (H, ψ) 是 Frobenius 代数;

(vi) (H, t) 是 Frobenius 余代数;

(vii) S 既是反代数同态又是反余代数同态.

注记 4.2.8 (i) S 称为弱双 Frobenius 代数的对极.

(ii) 当 $\Delta(1) = 1 \otimes 1$ 时, 弱双 Frobenius 代数成为普通的双 Frobenius 代数.

下面我们给出几个弱双 Frobenius 代数的例子.

例 4.2.9　设 H 为有限维代数和余代数, 且满足弱双 Frobenius 代数定义条件 (i) 到条件 (iv), 假设存在 $\psi \in H^*$ 和 $t \in H$, 满足 $S : H \longrightarrow H, h \longrightarrow \sum \psi(t_1 h)t_2$, 且既是反代数同态又是反余代数同态, 则 (H, ψ, t, S) 是弱双 Frobenius 代数.

例 4.2.10　设 H 为有限维 Hopf 代数, (ψ, t) 为非退化右积分对 (即两者 $\psi \in H^*, t \in H$ 都是非退化右积分, 且满足 $t \leftharpoonup \psi = 1, \psi \leftharpoonup t = \varepsilon$), 则可得 $\sum \psi(hg_1)S(g_2) = \sum \psi(h_1 g)h_2$, 任取 $h, g \in H$.

令 $h = t$, 可得: $\sum (\psi \leftharpoonup t)(g_1) S(g_2) = \sum \psi(t_1 g) t_2$. 即 $S(g) = \sum \psi(t_1 g) t_2$, 任取 $g \in H$. 则我们得到 (H, ψ, t, s) 为弱双 Frobenius 代数.

例 4.2.11　设 (H, ψ_H, t_H, S_H) 和 (L, ψ_L, t_L, S_L) 均为弱双 frobenius 代数, 则 $(H \otimes L, \psi_H \otimes \psi_L, t_H \otimes t_L, S_H \otimes S_L)$ 也是弱双 Frobenius 代数.

更多关于弱双 Frobenius 代数的性质和例子, 见参考文献 [41].

4.3　(弱)(余) 拟 Hopf 代数

为了解决某种方程的解, V.G.Drinfel'd[71] 引进了拟 Hopf 代数 (quasi-Hopf algebra) 的概念, 随后有许多文献 [30, 147] 进行研究与讨论, 这也是我们本节介绍的主要内容.

定义 4.3.1　一个拟双代数(quasi-bialgebra)是一个四元组 $(H, \Delta, \varepsilon, \phi)$, 其中 H 是带有单位元 1_H 的结合代数, ϕ 是 $H \otimes H \otimes H$ 中可逆元素, $\Delta : H \longrightarrow H \otimes H$, $\varepsilon : H \longrightarrow k$ 是两个代数同态, 且满足下面的条件:

(i) $(i_d \otimes \Delta)\Delta(h) = \phi(\Delta \otimes i_d)\Delta(h)\phi$;

(ii) $(i_d \otimes \varepsilon)\Delta(h) = h = (\varepsilon \otimes i_d)\Delta(h)$,

对于所有的 $h \in H$, ϕ 是正则 3-余循环, 即 $(1_H \otimes \phi)(i_d \otimes \Delta \otimes i_d)(\phi)(\phi \otimes 1_H) = (i_d \otimes i_d \otimes \Delta)(\phi)(\Delta \otimes i_d \otimes i_d)(\phi)$ 和 $(i_d \otimes \varepsilon \otimes i_d)(\phi) = 1_H \otimes 1_H$.

采用 Sweedler 符号: 对于任取 $h \in H$, 记 $\Delta(h) = h_1 \otimes h_2$. 因为余乘法 Δ 知识满足拟结合律, 采用下列记号: $(\Delta \otimes i_d)\Delta(h) = h_{11} \otimes h_{12} \otimes h_2$ 和 $(i_d \otimes \Delta)\Delta(h) = h_1 \otimes h_{21} \otimes h_{22}$.

为了方便, 分别使用大写字母和小写字母来表示 ϕ 和 ϕ^{-1} 的分量, 即
$$\phi = X^1 \otimes X^2 \otimes X^3 = T^1 \otimes T^2 \otimes T^3 = V^1 \otimes V^2 \otimes V^3 = \cdots,$$
$$\phi^{-1} = x^1 \otimes x^2 \otimes x^3 = t^1 \otimes t^2 \otimes t^3 = v^1 \otimes v^2 \otimes v^3 = \cdots.$$

定义 4.3.2　我们称一个拟双代数为一个**拟-Hopf 代数**(quasi-Hopf algebra), 如果存在一个代数反同态 $S : H \longrightarrow H$ 和元素 $\alpha, \beta \in H$, 满足下列条件: 对于任意 $h \in H$, 有
$$S(h_1)\alpha h_2 = \varepsilon(h)\alpha, \quad h_1 \beta S(h_2) = \varepsilon(h)\beta,$$

4.3 (弱)(余) 拟 Hopf 代数

$$X^1\beta S(X^2)\alpha X^3 = 1 = S(x^1)\alpha x^2\beta S(x^3).$$

关于拟 Hopf 代数, 我们给出一个例子如下.

例 4.3.3 已知 G 是一个有限群, $C[G]$ 是其在复数域上的群代数, $F[G]$ 是定义在 G 上的复值函数作成的代数, 令 $c: G\otimes G\otimes G \longrightarrow \pi$ 是定义在 G 上的 3-余循环, 其取值在循环群 π 上. 这就意味着:

$$c(y,z,w)c(xy,z,w)^{-1}c(x,yz,w)x(x,y,zw)^{-1}x(x,y,z) = 1,$$

任取 $x,y,z,w \in G$. 我们假设 c 是正则的, 即 $c(x,y,z) = 1$ 当 x,y 或者 z 等于 G 中单位元 e. 我们定义一个拟 Hopf 代数 $D^C(G)$ 如下:

作为一个向量空间, $D^C(G) = F(G) \otimes C(G)$, 它的一组基底是这样的元素组成的:

$\delta_g \otimes x$, 其中 $g,x \in G$, 其中 δ_g 表示狄拉克三角函数, 即 $\delta_g(h) = 1$, 当 $h = g$. 否则 $\delta_g(h) = 0$. 令

$$\theta_g(x,y) = \frac{c(g,x,y)c(x,y,(xy)^{-1}gxy)}{c(x,x^{-1}gx,y)},$$

$$\gamma_x(g,h) = \frac{c(g,h,x)c(x,x^{-1}gx,x^{-1}hx)}{c(g,x,x^{-1}hx)}.$$

$D^c(G)$ 上面的乘法定义是

$$(\delta_g \otimes x)(\delta_h \otimes y) = \begin{cases} \theta_g(x,y)(\delta_g \otimes xy), & gx = xh, \\ 0, & \text{其他情况}, \end{cases}$$

单位元素是 $1 = \sum_g(\delta_g \otimes e)$, 余乘法、对极、余单位的定义分别是

$$\begin{cases} \Delta(\delta_g \otimes x) = \sum_{\{h,k\in G | hk=g\}} \gamma_x(h,k)(\delta_h \otimes x) \otimes (\delta_k \otimes x), \\ S(\delta_g \otimes x) = \theta_{g^{-1}}(x,x^{-1})^{-1}\Upsilon_x(g,g^{-1})^{-1}(\delta_{x^{-1}gx} \otimes x^{-1}), \\ \varepsilon(\delta_g \otimes x) = \delta_{g,e}1. \end{cases}$$

最后, 有 $\Phi = \sum_{\{g,h,k\in G\}} c(g,h,k)(\delta_g \otimes e) \otimes (\delta_h \otimes e) \otimes (\delta_k \otimes e)$. 于是, $D^c(G)$ 是一个拟-Hopf 代数.

我们上面给出了拟双代数与拟 Hopf 代数的定义, 下面给出这两种结构的对偶.

定义 4.3.4 一个**余拟双代数**(coquasi bialgebra) 是一个六元组 $(H, m, \mu, \omega, \Delta, \varepsilon)$, 其中 (H, Δ, ε) 是一个余代数, 乘法: $m: H \otimes H \longrightarrow H$(记为 $m(h \otimes g) = hg$) 单位 $\mu: k \longrightarrow H$ (记为 $\mu(1) = 1_H$) 是余代数同态, $\omega \in (H \otimes H \otimes H)^*$ 是一个卷积可逆元素, 使得:

(i) $h_1(g_1k_1)\omega(h_2, g_2, k_2) = \omega(h_1, g_1, k_1)(h_2g_2)k_2;$

(ii) $1_H h = h1_H = h$;

(iii) $\omega(h_1, g_1, k_1 l_1)\omega(h_2 g_2, k_2, l_2) = \omega(g_1, k_1, l_1)\omega(h_1, g_2 k_2, l_2)\omega(h_2, g_3, k_3)$;

(iv) $\omega(h, 1_H, g) = \varepsilon(h)\varepsilon(g)$,

对于任取 $h, g, k, l \in H$ 都成立.

定义 4.3.5 一个**余拟 Hopf 代数**(coquasi Hopf algebra)是一个余拟双代数定义了余代数反同态 $S : H \longrightarrow H$ (对极) 和元素 $\alpha, \beta \in H^*$, 使得任取 $h \in H$, 都有:

(i) $S(h_1)\alpha(h_2)h_3 = \alpha(h)1_H$;

(ii) $h_1\beta(h_2)S(h_3) = \beta(h)1_H$;

(iii) $\omega(h_1\beta(h_2), S(h_3), \alpha(h_4)h_5) = \omega^{(-1)}(S(h_1), \alpha(h_2)h_3\beta(h_4), S(h_5)) = \varepsilon(h)$.

我们下面给出一个余拟 Hopf 代数的例子: 已知 H 为有限维交换 Hopf 代数, 上面有对极 S 和 $\Phi \in H \otimes H \otimes H$, 一个正则的 3-余循环 (3-cocycle), 即 Φ 卷积可逆而且满足下面的条件:

(i) $(i_d \otimes i_d \otimes \Delta)(\Phi)(\Delta \otimes i_d \otimes i_d)(\Phi) = (i_d \otimes \Phi)(i_d \otimes \Delta \otimes i_d)(\Phi)(\Phi \otimes i_d)$;

(ii) $(i_d \otimes i_d \otimes \varepsilon)(\Phi) = (i_d \otimes \varepsilon \otimes i_d)(\Phi) = (\varepsilon \otimes i_d \otimes i_d)(\Phi) = 1 \otimes 1$.

假设 H 弱余作用于 C(余代数), 令 $\alpha \longrightarrow H \otimes H$ 是一个线性映射, 记为 $\alpha(c) = c' \otimes c''$, 则我们有一个交叉余积 $C \times_\alpha H$, 上面有乘法:

(iii) $\widetilde{\Delta}(c \times h) = \sum c_1 \times c_2^{(1)} c_3^1 h_1 \otimes c_2^{(0)} \times c_3^2 h_2$, 其中 $\rho(c) = \sum c^{(1)} \otimes c^{(0)} \in H \otimes C$ 是 C 上的一个左 H-余模结构映射, 任取 $c \in C, h \in H$.

若 C 是余交换的 H 是交换的, 那么 $(C \times H, \widetilde{\Delta}, \widetilde{\varepsilon} = \varepsilon_C \otimes \varepsilon_H)$ 是一个余交换余代数当且仅当下面的条件成立:

(iv) $\sum c_1^{(1)} c_2' \otimes c_1^{(0)'} \otimes c_1^{(0)''} c_2'' = \sum c_1' c_2 \otimes c_1'' c_2' \otimes c_2''$.

定义 $F \in H \otimes H \otimes H$ 为:

(v) $F = \sum \Phi' \Psi_3^3 X_2^2 \otimes \Phi^2 \Psi' S(\Psi_2^3) \Psi_4^3 X' S(X_1^2) X_3^2 \otimes \Phi^3 \Psi^2 S(\Psi_1^3) \Psi_3^3 X^3$, 其中 $\Phi = \sum \Phi^1 \otimes \Phi^2 \otimes \Phi^3 = \Psi$, $\Phi^{-1} = \sum X^1 \otimes X^2 \otimes X^3$.

$\alpha : H^* \longrightarrow H \otimes H$:

(vi) $f \longrightarrow \sum \langle f, F^1 \rangle F^2 \otimes F^3$.

所以我们可以构造一个交叉余积 $H^* \times H$, 记为 $H^* \diamond_\Phi H$, 上面的余乘法为:

(vii) $\widetilde{\Delta}(f \times h) = \sum f_1 \times f_2^{(1)} f_3^1 h_1 \otimes f_2^{(0)} \times f_3^2 h_2$.

定义 $G \in H \otimes H \otimes H$ 为:

(viii) $G = \sum \Phi^1 \Psi_2^2 X^1 \otimes \Phi^2 \Psi_2^3 X_2^3 \otimes \Phi^3 \Psi^1 S(\Psi_1^2) \Psi_3^2 S(\Psi_3^1) \Psi_3^3 X^2 S(X_1^3) X_3^3$.

$\tau : H^\star \diamond_\Phi H \otimes H^\star \diamond_\Phi H \otimes H^\star \diamond_\Phi H \longrightarrow k$:

(ix) $\tau(f \otimes x \otimes g \otimes y \otimes u \otimes z) = \langle f, X^1 \rangle \langle g, X^2 \rangle \langle u, X^3 \rangle \varepsilon(x)\varepsilon(y)\varepsilon(z)$, 任取 $x, y, z \in H, f, g, u \in H^\star$.

定义 α 和 β: $H^\star \diamond_\Phi H \longrightarrow k$ 为: $\alpha(f \otimes h) = \langle f, 1\rangle\varepsilon(h)$ 和 $\beta(f \otimes h) = \langle f, \phi^1 S(\phi^2)\phi^3\rangle\varepsilon(h)$.

$S: H^\star \diamond_\Phi H \longrightarrow H^\star \diamond_\Phi H$ 为:

(x)$\alpha(f \otimes h) = S^\star(f_3^{(0)}) \otimes S(f^1 h) f_1^{(2)'} S(f_1^{(2)'}) \sigma^{-1}(S^\star(f_2^{(0)}) \otimes f_3^{(0)})$.

其中 $\sigma: H^\star \otimes H^\star \longrightarrow H$ 为 $\sigma(h^\star \otimes g^\star) = \sum \langle h^\star, G^1\rangle\langle g^\star, G^2\rangle G^3$.

我们可以证明: 若 H 是一个有限维 Hopf 代数, 那么 $(H^\star \diamond_\Phi H, \widetilde{\Delta}, \widetilde{\varepsilon}, \tau, \alpha, \beta)$ 是一个余拟 Hopf 代数.

将前面的拟 Hopf 代数公理削弱后, 我们可以得到另一种类 Hopf 结构.

定义 4.3.6 $(A, \Delta, \varepsilon, \Phi)$ 是一个**弱拟双代数**(weak quasi-bialgebra)[273], 如果它满足下列条件:

$(A, \Delta, \varepsilon, \Phi)$ 是具有单位元 1 的结合代数, 不保持单位的代数映射 $\Delta: A \longrightarrow A \otimes A$, 代数映射 $\varepsilon: A \longrightarrow k$, 并且存在拟逆元素 $\Phi \in (i_d \otimes \Delta)\Delta(1)(A \otimes A \otimes A)(\Delta \otimes i_d)\Delta(1)$, 满足:

(i) $(i_d \otimes \Delta)\Delta(a) = \Phi((\Delta \otimes i_d)\Delta(a))\Phi^{-1}$;
(ii) $(\varepsilon \otimes i_d)\Delta(a) = a, (i_d \otimes \varepsilon)(\Delta(a)) = a$;
(iii) $(i_d \otimes i_d \otimes \Delta)(\Phi)(\Delta \otimes i_d \otimes i_d)(\Phi) = \Phi_{234}(i_d \otimes \Delta \otimes i_d)(\Phi)\Phi_{123}$;
(iv) $(i_d \otimes \varepsilon \otimes i_d)(\Phi) = \Delta(1)$.

其中 $\Phi_{123} = \Phi \otimes 1, \Phi_{234} = 1 \otimes \Phi$. 由于 Φ 是拟逆的, 可只有下列等式: $\Phi = \Phi(\Delta \otimes i_d)(\Delta(1)) = (i_d \otimes \Delta)(\Delta(1))\Phi, \Phi\Phi^{-1} = (i_d \otimes \Delta)(\Delta(1))$.

定义 4.3.7 我们说一个弱拟双代数 $(A, \Delta, \varepsilon, \Phi)$ 是**弱拟 Hopf 代数**(weak quasi-Hopf algebra), 如果存在 A 上的可逆代数反同态 $s: A \longrightarrow A$, 以及 A 上的元素 α, β 有

$$\begin{cases} \sum s(a_1)\alpha a_2 = \varepsilon(a)\alpha, & \sum a_1\beta s(a_2) = \varepsilon(a)\beta, \quad a \in A, \\ \sum X^1\beta S(X^2)\alpha X^3 = 1, & \sum s(x^1)\alpha x^2\beta s(x^3) = 1. \end{cases}$$

这里记 $\Phi = \sum X^1 \otimes X^2 \otimes X^3, \Phi^{-1} = \sum x_1 \otimes x_2 \otimes x_3$.

4.4 (弱) 拟 Hopf 群 (余) 代数

(弱) 拟 Hopf 群 (余) 代数的研究已经是国际该领域研究的热点问题之一, 相关的概念理论, 如 Hopf 群余代数[31, 58, 285, 286]、群余分次乘子 Hopf 代数[1, 197, 267, 268] 以及其他相关概念理论研究, 参考文献见 [69, 86—88, 146, 169, 207, 208, 223, 226, 228–230, 232, 237].

定义 4.4.1 数域 k 上的一个**群代数**(group-algebra)是一族 k-向量空间 $A = \{A_\alpha\}_{\alpha \in \pi}$, 上面定义了一族 k-线性映射 $m = \{m_{\alpha,\beta}: A_\alpha \otimes A_\beta \longrightarrow A_{\alpha\beta}\}_{\alpha,\beta \in \pi}$

(叫做乘法) 和一个线性映射 $\eta : k \longrightarrow A_1$(叫做单位), 使得 m 满足结合律, 即: 任取 $\alpha, \beta, \gamma \in \pi$, $m_{\alpha,\beta,\gamma}(m_{\alpha,\beta} \otimes \mathrm{id}_{A_\gamma}) = m_{\alpha,\beta\gamma}(\mathrm{id}_{A_\alpha} \otimes m_{\beta,\gamma})$, 同时, η 是满足单位性质: $m_{\alpha,1}(\mathrm{id}_{A_\alpha} \otimes \eta) = \mathrm{id}_{A_\alpha} = m_{1,\alpha}(\eta \otimes \mathrm{id}_{A_\alpha})$.

注记 4.4.2 显然可以看到 $(A_1, m_{1,1}, \eta)$ 就是通常意义下的代数.

定义 4.4.3 一个 **Hopf 群代数** (Hopf group algebra) 是一个群代数 $H = (\{H_\alpha\}_{\alpha \in \pi}, m, \eta)$ 使得:

(i) 每一个 H_α 都是一个 k-余代数, 上面有余乘法 Δ_α, 余单位 $\varepsilon_\alpha, \alpha \in \pi$;

(ii) $\eta : k \longrightarrow H_1$ 和 $m_{\alpha,\beta} : H_\alpha \otimes H_\beta \longrightarrow H_{\alpha\beta}$ 是余代数映射, $\alpha, \beta \in \pi$;

(iii) 存在一族 k-线性映射 $S = \{S_\alpha : H_\alpha \longrightarrow H_{\alpha^{-1}}\}_{\alpha \in \pi}$(叫做对极) 使得下面的等式成立:

$$m_{\alpha^{-1},\alpha}(S_\alpha \otimes \mathrm{id}_{H_\alpha})\Delta_\alpha = \eta\varepsilon_\alpha = m_{\alpha,\alpha^{-1}}(\mathrm{id}_{H_\alpha} \otimes S_\alpha)\Delta_\alpha, \quad \alpha \in \pi.$$

例 4.4.4 已知 π 是一个由 α 生成的二阶循环群, 即 $\pi = (1, \alpha), \alpha^2 = 1$. 令 H_1 是一个二阶的 k-空间, 其基底为 (h_0, h_2), 令 H_α 是另一个二阶的 k-空间, 其基底为 (h_1, h_3). 定义 k-线性映射如下:

$$m_{1,1} : H_1 \otimes H_1 \longrightarrow H_1, m_{1,1}(h_0 \otimes h_0) = m_{1,1}(h_2 \otimes h_2) = h_0,$$
$$m_{1,1}(h_0 \otimes h_2) = m_{1,1}(h_2 \otimes h_0) = h_2;$$
$$m_{\alpha,\alpha} : H_\alpha \otimes H_\alpha \longrightarrow H_1, m_{\alpha,\alpha}(h_1 \otimes h_3) = m_{\alpha,\alpha}(h_3 \otimes h_1) = h_0,$$
$$m_{\alpha,\alpha}(h_1 \otimes h_1) = m_{\alpha,\alpha}(h_3 \otimes h_3) = h_2;$$
$$m_{1,\alpha} : H_1 \otimes H_\alpha \longrightarrow H_\alpha, m_{1,\alpha}(h_0 \otimes h_1) = m_{1,\alpha}(h_2 \otimes h_3) = h_1,$$
$$m_{1,\alpha}(h_0 \otimes h_3) = m_{1,\alpha}(h_2 \otimes h_1) = h_3;$$
$$m_{\alpha,1} : H_\alpha \otimes H_1 \longrightarrow H_\alpha, m_{\alpha,1} = m_{1,\alpha}\tau_{H_\alpha,H_1}\eta \longrightarrow H_1,$$
$$\eta(\lambda) = \lambda h_0, \quad \lambda \in k.$$

此时我们可以知道 $H = (H_1, H_\alpha, m, \eta)$ 当 $h_0 = 1$ 时是一个群代数.

定义线性映射 $\Delta_1 : H_1 \longrightarrow H_1 \otimes H_1, \Delta(h_i) = h_i \otimes h_i, \eta_1 : H_1 \longrightarrow k, \varepsilon_1(h_i) = 1, i = 0, 2$. $\Delta_\alpha : H_\alpha \longrightarrow H_\alpha \otimes H_\alpha, \Delta(h_i) = h_i \otimes h_i$. $\varepsilon_\alpha : H_\alpha \longrightarrow k, \varepsilon_\alpha(h_i) = 1, i = 1, 3$. 对极 $S = S_1, S_\alpha$: $S_1 : H_1 \longrightarrow H_1, h_0 \mapsto h_0, h_2 \mapsto h_2; S_\alpha : H_\alpha \longrightarrow H_\alpha, h_1 \mapsto h_3, h_3 \mapsto h_1$.

此时的 H 成为一个 Hopf 群代数.

定义 4.4.5 一个**弱 Hopf 群代数**(weak Hopf group algebra)是一个群代数 $H = (\{H_\alpha\}, m, \eta)$, 在其上定义了一族 k-线性映射 $S = \{S_\alpha : H_\alpha \longrightarrow H_{\alpha^{-1}}\}_{\alpha,\beta \in \pi}$(叫做对极) 满足如下的性质;

4.4 (弱) 拟 Hopf 群 (余) 代数

(i) 每一个 H_α 都是一个余代数, 具有余乘 Δ_α 和余单位 ε_α.

(ii) 任取 $\alpha, \beta \in \pi, m_{\alpha,\beta}:$
$H_\alpha \otimes H_\beta \longrightarrow H_{\alpha\beta}$ 和 $\eta : k \longrightarrow H_1$ 满足下面的条件:

$$\Delta_{\alpha\beta} m_{\alpha,\beta} = (m_{\alpha,\beta} \otimes m_{\alpha,\beta})(\mathrm{id}_{H_\alpha} \otimes \tau_{H_\alpha, H_\beta} \otimes \mathrm{id}_{H_\beta})(\Delta_\alpha \otimes \Delta_\beta),$$

$$\Delta_1^2(1) = 1_{(1,1)} \otimes 1_{(2,1)} 1_{(1',1)} \otimes 1_{(2',1)} = 1_{(1,1)} \otimes 1_{(1',1)} 1_{(2,1)} \otimes 1_{(2',1)},$$

其中 $1 \in H_1, \Delta_1(1) = 1_{(1,1)} \otimes 1_{(2,1)} = 1_{(1',1)} \otimes 1_{(2',1)}, \Delta_1^2(1) = 1_1 \otimes 1_2 \otimes 1_3$,

$$\varepsilon_{\alpha\beta\gamma}(h_\alpha h_\beta h_\gamma) = \varepsilon_{\alpha\beta}(h_\alpha h_{\beta(2,\beta)}) \varepsilon_{\beta\gamma}(h_{\beta(1,\beta)} h_\gamma) = \varepsilon_{\alpha\beta}(h_\alpha h_{\beta(1,\beta)}) \varepsilon_{\beta\gamma}(h_{\beta(2\beta)} h_\gamma),$$

其中 $h_\alpha \in H_\alpha, h_\beta \in H_\beta, h_\gamma \in H_\gamma$.

(iii) $S = \{S_\alpha : H\alpha \longrightarrow H_{\alpha^{-1}}\}_{\alpha,\beta \in \pi}$ 符合下面的条件: 任取 $h_\alpha \in H_\alpha$,

$$\begin{cases} h_{\alpha(a,\alpha)} S_\alpha(h_{\alpha(2,\alpha)}) = \varepsilon_\alpha(1_{(1,1)} h_\alpha) 1_{(2,1)}; \\ S_\alpha(h_{\alpha(1,\alpha)}) h_{\alpha(2,\alpha)} = 1_{(1,1)} \varepsilon_\alpha(h_\alpha 1_{(2,1)}); \\ S_\alpha(h_{\alpha)(a,\alpha)(a,\alpha)}) h_{\alpha(1,\alpha)(2,\alpha)} S_\alpha(h_{\alpha(2,\alpha)}) = S_\alpha(h_\alpha). \end{cases}$$

例 4.4.6 已知 $(H, \Delta, \varepsilon, m, 1, S)$ 是一个弱 Hopf 代数, 对于任意的 $\alpha, \beta \in \pi$, 定义 $A_\alpha = H, \Delta_\alpha = \Delta, \varepsilon_\alpha = \varepsilon, m_{\alpha,\beta} = m, S_\alpha = S$. 很容易就可以证明 $A = \{A_\alpha\}_{\alpha \in \pi}$ 是一个弱 Hopf 群代数.

定义 4.4.7 域 k 上的一个**群余代数**是带有一族 k-线性映射

$\Delta = \{\Delta_{\alpha,\beta} : C_{\alpha\beta} \longrightarrow C_\alpha \otimes C_\beta\}_{\alpha,\beta \in \pi}$ (叫做余乘), 和 $\varepsilon : C_i \longrightarrow k$ (叫做余单位) 的一族 k-线性空间 $\{C_\alpha\}_{\alpha \in k}$, 且对于任意的 $\alpha, \beta, \gamma \in \pi$, 有

(i) $(\Delta_{\alpha,\beta} \otimes \mathrm{id}_{C_r}) \Delta_{\alpha\beta,\gamma} = (\mathrm{id}_{C_\alpha} \otimes \Delta_{\beta,\gamma}) \Delta_{\alpha,\beta\gamma}$;

(ii) $((\mathrm{id}_{C_\alpha}) \otimes \varepsilon) \Delta_{\alpha,i} = (\varepsilon \otimes \mathrm{id}_{C_\alpha}) \Delta_{i,\alpha} = \mathrm{id}_{C_\alpha}$.

我们采用 Sweedler 余乘记号, 即对于任意的 $\alpha, \beta \in \pi, c \in C_{\alpha\beta}, \Delta_{\alpha,\beta}(c) = \sum c_{(1,\alpha)} \otimes c_{(2,\alpha)}$.

注记 4.4.8 $(C_i, \Delta_{i,i}, \varepsilon)$ 是通常意义下的余代数.

定义 4.4.9 一个 **Hopf 群余代数**(Hopf group coalgebra)是一个群余代数 $H = (\{H_\alpha\}, \Delta, \varepsilon)$, 上面定义了一族 k-线性映射 (称为对极)$S = \{S_x : H_\alpha \longrightarrow H_{\alpha^{-1}}\}_{\alpha \in \pi}$ 满足下面的条件:

(i) 每一个 H_α 都是一个代数, 乘法 m_α, 单位元素 $1_\alpha \in H_\alpha$.

(ii) $\varepsilon : H_1 \longrightarrow k, \Delta_{\alpha,\beta} \longrightarrow H_\alpha \otimes H_\beta$(任取 $\alpha, \beta \in \pi$) 是代数同态.

(iii) 任取 $\alpha \in \pi, m_\alpha(S_{\alpha^{-1}} \otimes \mathrm{id}_{H_\alpha}) \Delta_{\alpha^{-1},\alpha} = \varepsilon 1_\alpha = m_\alpha(\mathrm{id}_{H_\alpha} \otimes S_{\alpha^{-1}}) \Delta_{\alpha,\alpha^{-1}}$.

我们将 Hopf 群余代数和弱 Hopf 代数的思想结合在一起, 又可得一种新的结构: **弱 Hopf 群余代数**.

定义 4.4.10 域 k 上的一个弱半群 H-余代数 (或弱半 Hopf 群余代数) 是指一族代数 $\{H_\alpha, m_\alpha, 1_\alpha\}_{\alpha\in\pi}$, 同时也是群余代数 $(H_\alpha, \Delta, \varepsilon)_{\alpha,\beta\in\pi}$ 且对任意 $\alpha,\beta\in\pi$ 满足下列条件:

(i) $\Delta(h_{\alpha\beta}g_{\alpha\beta}) = \Delta(h_{\alpha\beta})\Delta(g_{\alpha\beta}), \varepsilon(1_i) = 1_i$;

(ii) $\Delta^2(1_{\alpha\beta\gamma}) = \sum 1_{\alpha\beta\gamma1\alpha\beta1\alpha} \otimes 1_{\alpha\beta\gamma1\alpha\beta2\beta} \otimes 1_{\alpha\beta\gamma2\gamma}$
$= \sum 1_{\alpha\beta1\alpha} \otimes 1_{\beta\gamma1\beta}1_{\alpha\beta2\beta} \otimes 1_{\beta\gamma2\gamma}$
$= \sum 1_{\alpha\beta1\alpha} \otimes 1_{\alpha\beta2\beta}1_{\beta\gamma1\beta} \otimes 1_{\beta\gamma2\gamma}$;

(iii) $\varepsilon(x_i y_i z_i) = \sum \varepsilon(x_i y_i 1)\varepsilon(y_{i2} z_i) = \sum \varepsilon(x_i y_{i2})\varepsilon(y_{i_1} z_i)$.

另外, 如果它还带有一族 k-线性映射 $S = \{S_\alpha : H_{\alpha^{-1}} \longrightarrow H_\alpha\}_{\alpha\in\pi}$(称为对极) 满足下列条件:

(iv) $\sum x_{i1\alpha} S_\alpha(x_{x2\alpha^{-1}}) = \varepsilon(1_{\alpha 1 i})1_{\alpha 2\alpha}$;

(v) $S_{\alpha^{-1}}(x_{i1\alpha})x_{i2\alpha^{-1}} = \sum 1_{\alpha^{-1}1\alpha^{-1}}\varepsilon(x1_{\alpha^{-1}2i})$;

(vi) $\sum S_{\alpha^{-1}}(x_{\alpha 1 \alpha})x_{\alpha 2 i 1 \alpha^{-1}}S_{\alpha^{-1}}(x_{\alpha 2 i 2 \alpha}) = S_{\alpha^{-1}}(x_\alpha)$,

则称其为**弱 Hopf 群余代数**(weak Hopf group coalgebra).

下面我们看一下弱 Hopf 群余代数的例子.

例 4.4.11 令 H 是带有余乘法 Δ^H, 余单位 ε^H, 对极 S^H 的弱 Hopf 代数, 对于任意的 $\alpha, \beta \in \pi$, 令 $C_\alpha = H, \Delta_{\alpha,\beta} = \Delta^H, \varepsilon = \varepsilon^H, S_\alpha = S^H$, 易证 $C = \{C_\alpha\}_{\alpha\in\pi}$ 是一个弱 Hopf 群余代数.

例 4.4.12 令 H 是带有余乘法 Δ^H、余单位 ε^H、对极 S^H 的弱 Hopf 代数, $A^\pi = \{A_\alpha\}_{\alpha\in\pi}$, 其中对于每一个 $\alpha \in \pi$ 代数 A_α 是 H 的复制. 固定一个代数同构: $\delta_\alpha(H \longrightarrow A_\alpha)$, 对于任意的 $\alpha, \beta \in \pi$, 我们定义余乘法: $\Delta_{\alpha,\beta} : A_{\alpha,\beta} \longrightarrow A_\alpha \otimes A_\beta$ 为: $\Delta_{\alpha,\beta}(\delta_{\alpha,\beta}(h)) = \sum \delta_\alpha(h_1) \otimes \delta_\beta(h_2)$, 其中 $\Delta(h) = h_1 \otimes h_2$ 是给定的 H 中的余乘法.

余单位为 $\varepsilon : A_i \longrightarrow k, \varepsilon(\delta_i(h)) = \varepsilon(h)$.

对极 $S = \{S_\alpha : A_{\alpha^{-1}} \longrightarrow A_\alpha\}_{\alpha\in\pi}, S_\alpha(\delta_{\alpha^{-1}}(h)) = \delta_\alpha(S(h))$.

不难验证此时的 $A^\pi = \{A_\alpha\}_{\alpha\in\pi}$ 是弱 Hopf 群余代数.

注记 4.4.13 目前最广的弱 Hom-Hopf 群余代数见文献 [269], 主要用于构造辫子交叉张量范畴 [289], 其他代数结构性质有待研究.

4.5 (弱)Hom-Hopf 代数

本节主要介绍 (弱)Hom-Hopf 代数的概念, 参考文献见 [77, 116, 260, 269, 270, 275].

定义 4.5.1 一个**Hom-结合代数** (Hom-associate algebra) 是一个四元组 (A, μ, η, α), 其中 A 是一个线性空间, $\alpha : A \longrightarrow A, \mu : A \otimes A \longrightarrow A, \eta : k \longrightarrow A$ 是

4.5 (弱)Hom-Hopf 代数

线性映射，记 $\mu(a \otimes b) = ab$, $\eta(1_k) = 1_A$，满足下面的条件：

(i) $\alpha(ab) = \alpha(a)\alpha(b)$;

(ii) $\alpha(a)(bc) = (ab)\alpha(c)$;

(iii) $\alpha(1_A) = 1_A$;

(iv) $1_A a = a 1_A = \alpha(a)$.

已知 A, B 是两个 Hom-代数，线性映射 $f : A \longrightarrow B$ 是一个 Hom-代数同态，当 $\alpha_B \circ f = f \circ \alpha_A$, $f(1_A) = 1_B$, $\mu_B \circ (f \otimes f) = f \circ \mu_A$.

对偶地，我们有如下定义.

定义 4.5.2 一个**Hom-余结合余代数**(Hom-coassociate coalgebra)是一个四元组 $(C, \Delta, \varepsilon, \alpha)$，其中 C 是一个线性空间，$\alpha : C \longrightarrow C$, $\Delta : C \longrightarrow C \otimes C$, $\varepsilon : C \longrightarrow k$ 是线性映射，记 $\Delta(c) = c_1 \otimes c_2$，满足下面的条件：

(i) $\Delta(\alpha(c)) = \alpha(c_1)\alpha(c_2)$;

(ii) $\alpha(c_1) \otimes \Delta(c_2) = \Delta(c_1) \otimes \alpha(c_2)$;

(iii) $\varepsilon \circ \alpha = \varepsilon$;

(iv) $\varepsilon(c_1)c_2 = c_1\varepsilon(c_2) = \alpha(c)$.

已知 C, D 是两个 Hom-余结合余代数，则线性映射 $f : C \longrightarrow D$ 称为 Hom-余代数同态，当 $\alpha_D \circ f = f \circ \alpha C$, $\varepsilon_c = \varepsilon_D \circ f$, $\Delta_D \circ f = (f \otimes f) \circ \Delta_C$.

定义 4.5.3 一个**Hom-双代数**(Hom-bialgebra)是一个六元组 $(H, \mu, \eta, \Delta, \varepsilon, \alpha)$，满足下面的条件：

(i) (H, α) 是 Hom-结合代数；

(ii) (H, α) 是 Hom-余结合余代数；

(iii) Δ, ε 是 Hom-代数同态.

定义 4.5.4 一个**Hom-Hopf 代数**(Hom-Hopf algebra)是一个七元组 $(H, \mu, \eta, \Delta, \varepsilon, S, \alpha)$，其中 (H, α) 是 Hom-双代数，$S : H \longrightarrow H$ 是线性映射满足下面的条件：

(i) $S \circ \alpha = \alpha \circ S$;

(ii) $a_1 S(a_2) = S(a_1)a_2 = \varepsilon(a)1_H$;

(iii) S 是 Hom-代数反同态和 Hom-余代数反同态.

我们进一步给出一种将 Hom-Hopf 代数削弱后的类 Hopf 代数.

定义 4.5.5 一个**弱 Hom-双代数** (weak Hom-bialgebra) 是一个六元组 $(H, \alpha, \mu, \eta, \Delta, \varepsilon)$，其中 (H, α) 是 Hom-代数和 Hom-余代数，满足下面的条件：

(i) $\Delta(ab) = \Delta(a)\Delta(b)$;

(ii) $\varepsilon((ab)c) = \varepsilon(ab_1)\varepsilon(b_2c)$, $\varepsilon(a(bc)) = \varepsilon(ab_2)\varepsilon(b_1c)$;

(iii) $(\Delta \otimes id)\Delta(1_H) = 1_1 \otimes 1_2 1'_1 \otimes 1'_2$, $(id \otimes \Delta)\Delta(1_H) = 1_1 \otimes 1'_1 1_2 \otimes 1'_2$.

定义 4.5.6　一个**弱 Hom-Hopf 代数**(weak Hom-Hopf algebra)是一个弱 Hom-双代数, 上面存在一个 k-线性映射 S(称为对极) 使得下面的条件成立:

(i) $S \circ \alpha = \alpha \circ S$;

(ii) $h_1 S(h_2) = \varepsilon_t(h)$, $S(h_1)h_2 = \varepsilon_s(h)$;

(iii) $S(hg) = S(g)S(h)$, $S(1_H) = 1_H$;

(iv) $\Delta(S(h)) = S(h_2) \otimes S(h_1)$, $\varepsilon \circ S = \varepsilon$.

下面我们给出几个例子.

例 4.5.7 (二维弱 Hom-Hopf 代数)　令 H_2 是以 I 为基的 k-空间, 我们在 H_2 上面定义如下的结构:

(1) 乘法的定义为

H_2	I	E
I	I	$I-E$
E	$I-E$	$I-E$

(2) 余乘法的定义为: $\Delta(I) = (I-E) \otimes (I-E) + E \otimes E, \Delta(E) = (I-E) \otimes (I-E)$;

(3) 余单位的定义为: $\varepsilon(I) = 2, \varepsilon(E) = 1$;

(4) α 的定义为: $\alpha(I) = I, \alpha(E) = I - E$;

(5) 反对极 S 的定义为: $S = \mathbf{id}_{H_2} : H_2 \longrightarrow H_2$.

不难直接计算可知 H_2 是一个弱 Hom-Hopf 代数.

例 4.5.8　(Sweedler 的五维弱 Hom-Hopf 代数)　设 H_4 以 $1_H, e, c, x, y$ 生成的向量空间, 下面定义如下结构:

(1) 乘法的定义为

H	1_H	e	c	x	y
1_H	1_H	e	c	x	y
e	e	e	c	x	y
c	c	c	e	$-y$	$-x$
x	x	x	y	0	0
y	y	y	x	0	0

(2) 余乘法的定义为

$$\Delta(1_H) = 1_H \otimes 1_H - 1_H \otimes e - e \otimes 1_H + 2e \otimes e,$$

$$\Delta(e) = e \otimes e, \quad \Delta(x) = c \otimes x + x \otimes e,$$

$$\Delta(c) = c \otimes c, \quad \Delta(y) = e \otimes y + y \otimes c.$$

(3) 余单位的定义为: $\varepsilon(1) = 2, \varepsilon(e) = \varepsilon(c) = 1, \varepsilon(x) = \varepsilon(y) = 0$.

(4) 定义 α: 在一组基 $(1_H, e, c, x, y)$ 下矩阵为

$$\begin{pmatrix} 1 & 0 & 0 & 0 & 0 \\ 0 & 1 & 0 & 0 & 0 \\ 0 & 0 & 1 & 0 & 0 \\ 0 & 0 & 0 & \lambda & 0 \\ 0 & 0 & 0 & 0 & \lambda \end{pmatrix} \quad (\text{其中 } 0 \neq \lambda \in k) \tag{4.5.1}$$

的线性变换.

(5) 反对极的定义 $S : H \longrightarrow H$: $S(1_H) = 1_H$, $S(e) = e$, $S(c) = c$, $S(x) = y, S(y) = -x$.

那么我们可以证明此时的 H_4 是弱 Hom-Hopf 代数.

4.6 (弱)Hopf (余) 拟群

本节主要介绍 (弱)Hopf (余) 拟群的概念, 参考文献见 [7, 8, 78, 100].

定义 4.6.1 设 H 是一个不一定结合但有单位的代数, 如果存在代数同态 $\Delta : H \to H \otimes H$ 和 $\varepsilon : H \to K$ 使得 H 成为一个余结合、余单位的余代数, 并且对任意 $h, g \in H$, 满足

$$Sh_1(h_2 g) = h_1(Sh_2 g) = \varepsilon(h)g,$$

$$(gSh_1)h_2 = (gh_1)Sh_2 = \varepsilon(h)g.$$

我们说 H 是一个**Hopf 拟群**(Hopf quasigroup).

例 4.6.2 一个 Hopf 代数本身就是一个 Hopf 拟群.

例 4.6.3 如果 G 是一个拟群, 定义乘法和单位都是 G 中的乘法和单位; 对任意的 $g \in G$, 若 $\Delta h = h \otimes h, \varepsilon(h) = 1, Sh = h^{-1}$, 那么 $H = KG$ 是一个 Hopf 拟群.

例 4.6.4 设 R 是一个交换环, L 是一个 I.P. loop, 有

$$RL = \oplus_{u \in L} Ru$$

是一个余交换的 Hopf 拟群. 它的乘法和单位都是 L 中的乘法和单位, 余乘、余单位、对极分别定义如下

$$\Delta_{RL} u = u \otimes u, \quad \varepsilon(u) = 1_R, \quad S_{RL} u = u^{-1},$$

在这种情况下, S_{RL} 是一个同构, 并且 $S_{RL} \circ S_{RL} = \mathbf{id}_{RL}$.

对偶地, 我们有如下定义.

定义 4.6.5 设 H 是一个不一定余结合但有余单位的余代数, 如果存在代数同态 $m: H \otimes H \to H$ 和 $u: K \to H$ 使得 H 成为一个单位结合代数, 并且对任意 $h \in H$, 满足

$$Sh_1h_{21} \otimes h_{22} = h_1 Sh_{21} \otimes h_{22} = 1 \otimes h,$$

$$h_1 1 \otimes Sh_{12}h_2 = h_1 1 \otimes h_{12}Sh_2 = h \otimes 1,$$

我们说 H 是一个 **Hopf 余拟群**(Hopf coquasigroup).

例 4.6.6 一个 Hopf 代数本身就是一个 Hopf 余拟群.

例 4.6.7 $A = K[S^{2^n-1}]$ 是一个 Hopf 余拟群, 对任意 $x_c \in A$, 定义结构映射如下:

$$\Delta x_c = \Sigma_{a+b=c} x_a \otimes x_b F(a,b),$$
$$\varepsilon(x_a) = \delta_{a,0},$$
$$Sx_a = x_a F(a,a).$$

例 4.6.8 $K[S^7]_{\rtimes}\mathbb{Z}_2^3$ 是一个非交换的 Hopf 余拟群.

例 4.6.9 $\mathbb{C}_q[S^3]$ 是一个魔方 Hopf 余拟群.

注记 4.6.10 一个有限维 Hopf 拟群的对偶是一个 Hopf 余拟群.

定义 4.6.11 设 (H, m, u) 是一个单位代数, (H, Δ, ε) 是一个余单位余代数, 如果对任意 $a, b, c \in H$, 满足:

(i) $\Delta(ab) = \Delta(a)\Delta(b)$.

(ii) $\varepsilon((ab)c) = \varepsilon(a(bc)) = \sum \varepsilon(ab_1)\varepsilon(b_2c) = \sum \varepsilon(ab_2)\varepsilon(b_1c)$.

(iii) 设 $\Delta 1_H = \sum x \otimes y \in H \otimes H$, 则 $\sum x_1 \otimes x_2 \otimes y = \sum x \otimes yx \otimes y = \sum x \otimes xy \otimes y$.

(iv) 如果我们如下定义 $\prod_H^L$ 和 $\prod_H^R$:

$$\prod_H^L(h) = \sum y\varepsilon(xh), \quad \prod_H^R(h) = \sum x\varepsilon(hy),$$

并且存在一个称为反对极的同态 $S: H \to H$ 使得下面的等式成立:

(a) $\sum h_1 Sh_2 = \prod_H^L(h) = \sum y\varepsilon(xh)$;

(b) $Sh_1 h_2 = \prod_H^R(h) = \sum x\varepsilon(hy)$;

(c) $\sum Sh_1 y\varepsilon(xh_2) = \sum x Sh_2 \varepsilon(h_1y) = Sh$;

(d) $\sum Sa_1(a_2b) = \sum xb\varepsilon(ay)$;

(e) $\sum a_1(Sa_2b) = \sum yb\varepsilon(xa)$;

(f) $\sum (ab_1)Sb_2 = \sum ay\varepsilon(xb)$;

(g) $\sum (aSb_1)b_2 = \sum ax\varepsilon(by)$,

则我们称 H 是一个**弱 Hopf 拟群**(weak Hopf quasigroup)[7].

例 4.6.12 一个弱 Hopf 代数本身就是一个弱 Hopf 拟群.

例 4.6.13 设 (H, m, u) 是一个单位代数, (H, Δ, ε) 是一个余单位余代数, 并且定义 4.6.11 (a), (b), (c) 成立. 定义如下同态:

$H_L = \text{Im}(\Pi_H^L)$,

$P_L : H \to H_L; \quad i_L : H_L \to H; \quad i_L \circ P_L = \Pi_H^L; \quad P_L \circ i_L = \mathbf{id}_{H_L}$,

$q_L^1 : H \otimes H \to H \times_L^1 H, \ j_L^1 : H \times_L^1 H \to H \otimes H$,

$q_R^1 : H \otimes H \to H \times_R^1 H, \ j_R^1 : H \times_R^1 H \to H \otimes H$,

$q_L^2 : H \otimes H \to H \times_L^2 H, \ j_L^2 : H \times_L^2 H \to H \otimes H$,

$q_R^2 : H \otimes H \to H \times_R^2 H, \ j_R^2 : H \times_R^2 H \to H \otimes H$.

定义左、右 Galois 同态分别为

$$\gamma = (\mathbf{id}_H \otimes m_H) \circ (\Delta_H \otimes \mathbf{id}_H), \quad \beta = (m_H \otimes \mathbf{id}_H) \circ (\mathbf{id}_H \otimes \Delta_H).$$

又定义同态:

$$f = q_R^1 \circ \beta \circ j_L^1, \quad g = q_L^2 \circ \gamma \circ j_R^2.$$

如果 f, g 都是同构, 并且 $j_L^1 \circ f^{-1} \circ q_R^1$ 是几乎左 H-线性的, $j_R^2 \circ g^{-1} \circ q_L^2$ 是几乎右 H-线性的, 那么 H 就是一个弱 Hopf 拟群.

注记 4.6.14 最近的文章 [271, 272] 研究了如何用 Hopf (余) 拟群来构造新的辫子张量范畴.

4.7 Hopf 代数胚

本节主要内容来自文献 [26, 29].

定义 4.7.1 设 R 是一个单位元. 一个 R 上的**左双代数胚**(left bialgebroid) A 满足下面公理:

(i) 存在两个环同态 $s : R \longrightarrow A$(源映射) 和 $t : R^{\text{op}} \longrightarrow A$(靶映射) 使得 $s(r)t(r') = t(r')s(r)$ 对任意 $r, r' \in R$, 因此, A 构成了 R-R-双模, 结构为

$$r \cdot a \cdot r' := s(r)t(r')a, \quad \text{对任意 } a \in A, r, r' \in R.$$

(ii) R-R-双模映射 $\Delta : A \longrightarrow A \otimes_R A$ 和 $\varepsilon : A \longrightarrow R$ 使得 (A, Δ, ε) 构成双模范畴 $_R\mathbb{M}_R$ 中的余半群 (comonoid), 即 R-余环 (coring).

(iii) 在下面意义下, Δ 保持乘法, 一般地, 尽管 $A \otimes_R A$ 没有环结构, 但是它的子双模

$$A \times^l_R A = \{X \in A \otimes_R A \mid X(t(r) \otimes 1) = X(1 \otimes s(r)), \forall r \in R\}$$

是一个环, 乘法定义为: $(a \otimes b)(a' \otimes b') = (aa' \otimes bb')$ 对任意 $a, a', b, b' \in A$. 现在, 要求 $\Delta : A \longrightarrow A \times^l_R A$ 是一个环同态, 即满足: $\Delta(ab) = \Delta(a)\Delta(b)$, $\Delta(1) = 1 \otimes 1$ 和 $\Delta(a)(t(r) \otimes 1) = \Delta(a)(1 \otimes s(r))$.

(iv) ε 保持单元性: $\varepsilon(1_A) = 1_R$.

(v) 在下面意义下, ε 和乘法相容:

$$\varepsilon(as(\varepsilon(b))) = \varepsilon(ab) = \varepsilon(at(\varepsilon(b))).$$

定义 4.7.2 设 R 是一个单位元. 一个 R 上的**右双代数胚** A 满足下面公理:

(i) 一个环 A 和两个环同态 $s : R \longrightarrow A$ 和 $t : R^{op} \longrightarrow A$ 使得 $s(r)t(r') = t(r')s(r)$ 对任意 $r, r' \in R$, 因此, A 构成了 R-R-双模, 结构为

$$r \cdot a \cdot r' := at(r)s(r'), \quad \text{对任意} \, a \in A, r, r' \in R.$$

(ii) R-R-双模映射 $\Delta : A \longrightarrow A \otimes_R A$ 和 $\varepsilon : A \longrightarrow R$ 使得 (A, Δ, ε) 构成双模范畴 $_R\mathbb{M}_R$ 中的余半群, 即 R-余环.

(iii) 在下面意义下, Δ 保持乘法: $\Delta(ab) = \Delta(a)\Delta(b)$, $\Delta(1) = 1 \otimes 1$, 一般地, 尽管 $A \otimes_R A$ 没有环结构, 但是它的子双模

$$A \times^r_R A = \{X \in A \otimes_R A \mid (s(r) \otimes 1)X = (1 \otimes t(r))X, \forall r \in R\}$$

是一个环, 乘法定义为: $(a \otimes b)(a' \otimes b') = (aa' \otimes bb')$ 对任意 $a, a', b, b' \in A$. 现在, 要求 $\Delta : A \longrightarrow A \times^r_R A$ 是一个环同态, 即满足: $\Delta(ab) = \Delta(a)\Delta(b)$ 和 $(s(r) \otimes 1)\Delta(a) = (1 \otimes t(r))\Delta(a)$.

(iv) ε 保持单元性: $\varepsilon(1_A) = 1_R$.

(v) 在下面意义下, ε 和乘法相容:

$$\varepsilon(t(\varepsilon(a))b) = \varepsilon(ab) = \varepsilon(s(a)b).$$

注记 4.7.3 (i) 一个左双代数胚 $(A, R, s, t, \Delta, \varepsilon)$ 也称为 Takeuchi $\times_L$-双代数. 这里 A 和 R 是结合单位环, 分别称为**全环**(total ring) 和**基环**(base ring).

(ii) 设 $A_L = (A, R, s, t, \Delta, \varepsilon)$ 是一个左双代数胚, 那么 $(A, R^{op}, t, s, \Delta^{cop}, \varepsilon)$ 和 $(A^{op}, R, t, s, \Delta, \varepsilon)$ 是一个右双代数胚.

4.7 Hopf 代数胚

(iii) 设 $\mathcal{A}_L = (A, R, s, t, \Delta, \varepsilon)$ 是一个左双代数胚, 那么 $\Delta(s(r)) = s(r) \otimes 1$ 和 $\Delta(t(r)) = 1 \otimes t(r)$ 对任意 $r \in R$.

(iv) 一个左双代数胚同态 (构): $\mathcal{A}_L = (A, L, s_l, t_l, \Delta_l, \varepsilon_l) \longrightarrow \mathcal{A}'_{L'} = (A', L', s'_l, t'_l, \Delta'_l, \varepsilon'_l)$ 是一对环同态 (构) $(\Phi : A \longrightarrow A', \varphi : L \longrightarrow L')$ 使得下面等式成立:

$$s'_l \circ \varphi = \Phi \circ s_l, \quad t'_l \circ \varphi = \Phi \circ t_l, \quad \varepsilon'_l \circ \Phi = \varphi \circ \varepsilon_l, \quad \Delta'_l \circ \Phi = (\Phi \otimes \Phi) \circ \Delta_l.$$

(v) 一个左 (右) 双代数胚也可以等价描述为: 设 $R^{\mathrm{op}} \xrightarrow{t} A \xleftarrow{s} R$ 是环范畴中图, 使得 R 在 A 上的左、右作用为: $r \cdot a := at(r), a \cdot r := as(r)$, 这使得 A 构成双 (R, R)-模, 等价地说, 我们要求 s 与 t 的像可以交换. 那么, 环 R 和双模 A 连同 A 在 $_A\mathbb{M}(\mathbb{M}_A)$ 中幺半群结构称为一个左 (右) 双代数胚当且仅当遗忘函子: $_A\mathbb{M} \longrightarrow {}_R\mathbb{M}_R (\mathbb{M}_A \longrightarrow {}_R\mathbb{M}_R)$ 是强 (或严格) 张量函子.

因此, 关于一个左双代数胚 $\mathcal{A}_L = (A, R, s, t, \Delta, \varepsilon)$ 上的 "模代数" 的自然候选是 A-模范畴中的幺半群.

详细地说, 一个左双代数胚 $\mathcal{A}_L = (A, R, s, t, \Delta, \varepsilon)$ 上的一个**左 A-模代数胚 M** 满足下面条件:

(v1) M 是一个左 A-模, 由此自然有一个双 R-模结构:

$$r \cdot m \cdot r' = (r \cdot 1_A \cdot r') \triangleright m = s(r)t(r') \triangleright m, \text{ 对任意 } m \in M, r, r' \in R;$$

(v2) 存在一个结合乘法 $\mu_M : M \otimes_R M \longrightarrow M, m \otimes n \mapsto mn$, 满足

$$a \triangleright (mn) = (a_1 \triangleright m)(a_2 \triangleright n), \text{ 对任意 } a \in A, m, n \in M;$$

(v3) 对于乘法 μ_M, 存在一个单位 $\eta_M : R \longrightarrow M, r \mapsto r \cdot 1_M \equiv 1_M \cdot r$ 满足

$$a \triangleright 1_M = \varepsilon(a) \cdot 1_M, \text{ 对任意 } a \in A.$$

注意, 我们有等式: $(m \cdot r)n = m(r \cdot n)$, $r \cdot (mn) = (r \cdot m)n$ 和 $(mn) \cdot r = m(n \cdot r)$. 同时, 条件 (v2) 和 (v3) 表示一个事实: (M, μ_M, η_M) 不仅是范畴 $_R\mathbb{M}_R$ 中而且也是范畴 $_A\mathbb{M}$ 中的幺半群.

M 的不变子集合 (子环) 定义为

$$M^A := \{m \in M \mid a \triangleright m = s(\varepsilon(a)) \triangleright m, a \in A\}$$
$$= \{m \in M \mid a \triangleright m = t(\varepsilon(a)) \triangleright m, a \in A\}.$$

(vi) 类似地, 对于右双代数胚, 可以定义右模代数胚.

定义 4.7.4 一个 Hopf 代数胚是一对 $(\mathcal{A}_L, S)$, 这里 $\mathcal{A}_L = (A, L, s, t, \Delta, \varepsilon)$ 是一个左双代数胚和 $S : A \longrightarrow A$ 是一个反自同构, 满足下面条件:

(i) $S \circ t = s,$ (4.7.1)

(ii) $S^{-1}(a_2)_1 \otimes S^{-1}(a_2)_2 a_1 = S^{-1}(a) \otimes 1_A,$ (4.7.2)

(iii) $S(a_1)_1 a_2 \otimes S(a_1)_2 = 1_A \otimes S(a)$ (4.7.3)

在 $A_L \otimes_L A$ 中, 对任意 $a \in A$.

注记 4.7.5 (i) 由 (4.7.3) 可推出：

$$S(a_1)a_2 = t(\varepsilon(S(a))).$$ (4.7.4)

(ii) 设 $\theta = \varepsilon \circ S \circ s : S : L \longrightarrow L$, 由 (4.7.4) 可推出：

$$t \circ \theta = S \circ s, \quad \theta(hl) = \theta(h)\theta(l),$$

由此, 有 $S : {}_R A_R \longrightarrow {}_L A_L$ 是一个扭曲双模映射, 这里 $R \cong L^{op}$ 作为环同构, A 的双 A-模结构如下: 对于固定的环同构 $\gamma : R \cong L^{op}$,

$$r \cdot a \cdot r' := as \circ \theta^{-1} \circ \gamma(r) t \circ \gamma(r').$$ (4.7.5)

事实上, (4.7.5) 使得 A 成为 $R \cong L^{op}$ 上的一个右双代数胚 (见命题 4.7.6 (ii)).

(iii) 我们有两个映射：

$$S_{A_L \otimes_L A} : A_L \otimes_L A \longrightarrow A_R \otimes_R A, \ a \otimes b \mapsto S(b) \otimes S(a),$$

$$S_{A_R \otimes_R A} : A_R \otimes_R A \longrightarrow A_L \otimes_L A, \ a \otimes b \mapsto S(b) \otimes S(a).$$

命题 4.7.6 设 $\mathcal{A}_L = (A, L, s_l, t_l, \Delta_l, \varepsilon_l)$, 那么下面等价：

(i) $(\mathcal{A}_L, S)$ 是一个 Hopf 代数胚.

(ii) $\mathcal{A}_L = (A, L, s_l, t_l, \Delta_l, \varepsilon_l)$ 是一个左双代数胚, $S : A \longrightarrow A$ 是一个反自同构, 满足条件 (4.7.1) 和 (4.7.4), 以及

$$S_{A_L \otimes_L A} \circ \Delta \circ S^{-1} = S_{A_R \otimes_R A}^{-1} \circ \Delta \circ S,$$ (4.7.6)

$$(\Delta_l \otimes \mathbf{id}_A) \circ \Delta_r = (\mathbf{id}_A \otimes \Delta_r) \circ \Delta_l : A \longrightarrow A_L \otimes_L A_R \otimes_R A,$$ (4.7.7)

$$(\Delta_r \otimes \mathbf{id}_A) \circ \Delta_l = (\mathbf{id}_A \otimes \Delta_l) \circ \Delta_r : A \longrightarrow A_R \otimes_R A_L \otimes_L A,$$ (4.7.8)

这里 R 和 ${}_R A_R$ 如 (2.7.5) 所定义, 且

$$\Delta_r = S_{A_L \otimes_L A} \circ \Delta \circ S^{-1} = S_{A_R \otimes_R A}^{-1} \circ \Delta \circ S : A \longrightarrow A_R \otimes_R A.$$

(iii) $\mathcal{A}_L = (A, L, s_l, t_l, \Delta_l, \varepsilon_l)$ 是一个左双代数胚, 且 $\mathcal{A}_R = (A, R, s_r, t_r, \Delta_r, \varepsilon_r)$

4.7 Hopf 代数胚

是一个右双代数胚使得 $R \cong L^{\mathrm{op}}$. S 是加群双射且

$$s_l(L) = t_r(R), \quad t_l(L) = s_r(R) \text{ 作为 } A \text{ 的子环,} \tag{4.7.9}$$

$$(\Delta_l \otimes \mathbf{id}_A) \circ \Delta_r = (\mathbf{id}_A \otimes \Delta_r) \circ \Delta_l, \quad (\Delta_r \otimes \mathbf{id}_A) \circ \Delta_l = (\mathbf{id}_A \otimes \Delta_l) \circ \Delta_r, \tag{4.7.10}$$

$$S(t_l(l)at_l(l')) = s_l(l')S(a)s_l(l), \quad S(t_r(r)at_r(r')) = s_r(r')S(a)s_r(r), \tag{4.7.11}$$

$$S(a_1)a_2 = s_r \circ \varepsilon_r(a), \quad a_1 S(a_2) = s_l \circ \varepsilon_l(a) \tag{4.7.12}$$

对任意 $a \in A, l, l' \in L, r, r' \in R$.

(iv) $\mathcal{A}_L = (A, L, s_l, t_l, \Delta_l, \varepsilon_l)$ 是一个左双代数胚, 且 $\mathcal{A}_R = (A, R, s_r, t_r, \Delta_r, \varepsilon_r)$ 是一个右双代数胚使得 $R \cong L^{\mathrm{op}}$, (4.7.9) 和 (4.7.10) 成立. 进一步, 加群同态

$$\alpha: A_R \otimes_R A \longrightarrow A_L \otimes_L A, a \otimes b \mapsto a_1 \otimes a_2 b, \quad \beta: A_R \otimes_R A \longrightarrow A_L \otimes_L A, a \otimes b \mapsto b_1 a \otimes b_2$$

是双射.

例 4.7.7 (i) **Hopf 代数** 如果 A 是交换环 R 上的一个代数, 且带有两个都等于单位映射的映射 $s = t = \mu: R \longrightarrow A$, 那么左或右 R-Hopf 代数胚就是 Hopf 代数结构.

(ii) **弱 Hopf 代数** 一个弱 Hopf 代数 H 带有双射反对极 S 是一个它的靶子代数 H_t(一个可分代数) 上的左双代数胚 (实际上, 是一个 H_t 上的 Hopf 代数胚). 反之, 如果 H 是一个可分代数 R 上的双代数胚, 那么 H 有一个弱 Hopf 代数结构.

具体来说: 令 $R = H_s$ 和 $L = H_t$, 那么 $S, S^{-1}: R \longrightarrow L$ 是代数反同构. 我们有四个 L, R 在 H 上的可换作用:

$$H_R: h \cdot r := hr, \quad {}_R H: r \cdot h := h S^{-1}(r),$$

$$H_L: h \cdot l := S^{-1}(l)h, \quad {}_L H: l \cdot h := lh.$$

设 $p_R: H \otimes H \longrightarrow H_R \otimes_R H$ 和 $p_L: H \otimes H \longrightarrow H_L \otimes_L H$ 是经典投射. 那么有一个左双代数胚: $\mathcal{H}_L = (H, L, \mathbf{id}_L, S^{-1}|_L, p_L \circ \Delta, \varepsilon_t)$ 和一个右双代数胚: $\mathcal{H}_R = (H, R, \mathbf{id}_R, S^{-1}|_R, p_R \circ \Delta, \varepsilon_s)$. 进一步, $(\mathcal{H}_L, \mathcal{H}_R, S)$ 满足命题 4.7.6(iii).

(iii) $H = A \otimes A^{\mathrm{op}}$. 设 A 是任意一个代数, $H = A \otimes A^{\mathrm{op}}$, 那么 H 是 A 上的 Hopf 代数胚, 具有下面结构:

源映射: $s: A \longrightarrow H, a \mapsto a \otimes 1$, 靶映射: $t: A \longrightarrow H, a \mapsto 1 \otimes a$;

余积: $\Delta: H \longrightarrow H \otimes_A H, a \otimes b \mapsto (a \otimes 1) \otimes (1 \otimes b)$;

余单位: $\varepsilon: H \longrightarrow A, a \otimes b \mapsto ab$;

反对极: $S: H \longrightarrow H, a \otimes b \mapsto b \otimes a$.

(iv) $H = \mathrm{End}(A)$. 设 A 是一个有限维代数, $H = \mathrm{End}(A)$, 那么 H 是 A 上的双代数胚, 具有下面结构:

源映射: $s: A \longrightarrow H, a \mapsto l_a$, 靶映射: $t: A \longrightarrow H, a \mapsto r_a$;

余积: $\Delta: H \longrightarrow H \otimes_A H \stackrel{\alpha}{\cong} \mathrm{Hom}_k(A \otimes A, A), h \mapsto \Delta(h)(a \otimes b) = h(ab)$, 这里 $\alpha(h \otimes l)(a \otimes b) = h(a)l(b)$;

余单位: $\varepsilon: H \longrightarrow A, h \mapsto h(1_A)$;

反对极: $S: H \longrightarrow H, a \otimes b \mapsto b \otimes a$.

命题 4.7.8 设 $\mathcal{A}_L = (A, L, s_l, t_l, \Delta_l, \varepsilon_l)$ 是一个左双代数胚, $S: A \longrightarrow A$ 是一个反自同构满足 (4.7.1) 和 (4.7.4). 假设 $\Delta_l = p_l \circ \Delta$, 这里 $p_l: A \otimes A \longrightarrow A_L \otimes_L A$ 是经典投射, $\Delta: A \longrightarrow A \otimes A$ 是余结合的 (可能是非余单位的) 余积满足:

$$p_l \circ (S \otimes S) \circ \Delta^{\mathrm{cop}} = p_l \circ \Delta \circ S, \quad p_l \circ (S^{-1} \otimes S^{-1}) \circ \Delta^{\mathrm{cop}} = p_l \circ \Delta \circ S^{-1}$$

那么 $(\mathcal{A}_L, S)$ 是一个 Hopf 代数胚.

例 4.7.9 (1) **群胚 Hopf 代数胚** 设 G 是一个群胚. 群胚代数 kG 有一个左双代数胚, 结构为: 若 G^0 有限, 则有单位元 $1 = \sum_{a \in G^0} a$, 基代数为 $L = kG^0$, $s_l = t_l$ 是自然嵌入映射, $\Delta_l(g) = g \otimes_L g, \varepsilon_l = t(g) = \mathrm{target}(g)$. 进一步, 它是一个 Hopf 代数胚, 具有反对极 $S(g) = g^{-1}$. 事实上, 这个例子满足命题 4.7.8 的条件, 带有 $\Delta(g) = g \otimes g$.

(2) **代数量子环面** 设 T_q 是由两个可逆元素 U, V 生成的域 k 上的单位结合代数, 满足 $UV = qVU$, 这里 q 是 k 中可逆元素. 设 L 是由元素 U 生成的 k 上的单位结合子代数, $s_l = t_l$ 是嵌入映射.

$$\Delta_l(U^n V^m) := U^n V^m \otimes_L V^m = V^m \otimes_L U^n V^m, \quad \varepsilon_l(U^n V^m) := U^n,$$
$$S(U^n V^m) := V^{-m} U^n.$$

经典投射 $p_l: T_q \otimes T_q \longrightarrow T_{qL} \otimes_L T_q$ 的横面 (section) ξ 具有如下定义:

$$\xi(U^n V^m{}_L \otimes_L U^k V^l) := U^{(n+k)} V^m \otimes V^l.$$

事实上, 这个例子满足命题 4.7.8 条件, 带有 $\Delta(U^n V^m) := U^n V^m \otimes V^m$.

(3) **冲积** 设 $(H, \Delta_H, \varepsilon_H, \tau)$ 是一个 Hopf 代数, 带有 τ 是双射. 设 $(L, \cdot, \rho) \in {}_H\mathcal{YD}^H$ 是一个辫子交换代数 (即 $ab = b_0(b_{(1)} \cdot a)$). 冲积代数 $L\#H$ 是 L 上的一个 Hopf 代数胚, 结构为

$$s_l(l) = l\#1_H, \quad t_l(l) = \rho(l) = l_0 \otimes l_{(1)},$$
$$\Delta_l(l\#h) = (l\#h_1) \otimes_L (1_L \otimes h_2), \quad \varepsilon_l(l\#h) = \varepsilon_H(h)l,$$
$$S(l\#h) = (\tau(h_2)\tau^2(l_{(1)})) \cdot l_0\#\tau(h_1)\tau^2(l_{(2)}).$$

最后, 简单地用环和余环的语言来定义 Hopf 代数胚.

4.7 Hopf 代数胚

定义 4.7.10 设 A 为代数, $A^e = A \otimes A^{op}$ 为其包络代数. 称 (H, i) 为 A^e-**环**(这里 $i: A^e \longrightarrow H$), 如果存在代数同态 $s: A \longrightarrow H$ 和反代数同态 $t: A \longrightarrow H$, 使得 $s(a)t(b) = t(b)s(a)$ 对任意 $a, b \in A$. 详细地, $s(a) = i(a \otimes 1), t(a) = i(1 \otimes a^{op})$, 反之, $i(a \otimes b^{op}) = s(a)t(b^{op})$. 称 A 为**基代数**; 称 H 为**全代数**; 称 s 为**源映射**; 称 t 为**靶映射**.

进一步, 对于一个 A^e-环 (H, s, t), 它也是一个 A^e-双模, 这意味着 H 是 A-双模: $a \cdot h \cdot b = s(a)hs(b)$; 同时, H 是 A^{op}-双模: $a^{op} \cdot h \cdot b^{op} = t(a^{op})ht(b^{op})$.

例 4.7.11 $\text{End}(A)$ 是一个标准的 A^e-环, 其中: $i: A^e \longrightarrow \text{End}(A), i(a \otimes b^{op})(x) = axb; s(a)(x) = ax, t(b)(x) = xb$. 进一步, 通过映射 i, 则 $\text{End}(A)$ 是一个 A^e-双模, 例如, $\text{End}(A)$ 是一个左 A^e-模:

$$(a \cdot f)(b) = af(b), \quad (a^{op} \cdot f)(b) = f(b)a^{op}$$

对任意 $a, b \in A, f \in \text{End}(A)$.

设 (H, s, t) 为 A^e-环, 那么 H 也可以看成 A-双模: 带有左作用和右作用:

$$a \cdot h = s(a)h, \quad h \cdot a = t(a)h, \text{ 对任意 } a \in A, h \in H. \tag{4.7.13}$$

考虑 Abelian 群 $H \otimes_A H$, 这是一个 A-双模: $a \cdot (g \otimes_A h) = gt(a) \otimes_A h, (g \otimes_A h) \cdot a = g \otimes_A hs(a)$, 对任意 $a \in A, g \otimes_A h \in H \otimes_A H$.

定义 $\Gamma = \Gamma(H, s, t) := (H \otimes_A H)^A$, 即

$$\Gamma = \left\{ g = \sum g^1 \otimes_A g^2 \in H \otimes_A H \,\Big|\, \sum g^1 t(a) \otimes_A g^2 = \sum g^1 \otimes_A g^2 s(a) \right\}.$$

命题 4.7.12 设 (H, s, t) 为 A^e-环, 那么 $\Gamma = \Gamma(H, s, t)$ 是一个 A^e-环, 其乘法: $(\sum g^1 \otimes_A g^2)(\sum h^1 \otimes_A h^2) = \sum g^1 h^1 \otimes_A g^2 h^2$, 单位:$1_H \otimes_A 1_H$, 代数同态 $i: A^e \longrightarrow \Gamma, a \otimes_A b^{op} \mapsto s(a) \otimes_A t(b^{op})$.

设 (H, s, t) 为 A^e-环, 通过公式 (4.7.13), H 看成 A-双模. 同时, $H \otimes_A H$ 也可以看成自然的 A-双模: $a \cdot (g \otimes_A h) \cdot b = s(a)g \otimes_A t(b)h$.

定义 4.7.13 设 (H, s, t) 为 A^e-环. 我们说 $(H, s, t, \Delta, \varepsilon)$ 是一个 A-**双代数胚**, 如果下面条件成立:

(B1) (H, Δ, ε) 是一个 A-余环;

(B2) $\text{Im}(\Delta) \subseteq \Gamma(H, s, t)$ 且 Δ 的余限制 (corestriction) $\Delta: H \longrightarrow \Gamma(H, s, t)$ 是一个代数同态;

(B3) $\varepsilon(1_H) = 1_A, \varepsilon(gh) = \varepsilon(gs(\varepsilon(h))) = \varepsilon(gt(\varepsilon(h)))$ 对任意 $g, h \in H$.

A-双代数胚 H 的反对极是一个反代数同态 $\tau: H \longrightarrow H$ 满足:

(ANT1) $\tau \circ t = s$;

(ANT2) $m_H \circ (\tau \otimes H) \circ \Delta = t \circ \varepsilon\tau$;

(ANT3) 存在一个自然投射 $H \otimes H \longrightarrow H \otimes_A H$ 的截面 $\gamma: H \otimes_A H \longrightarrow H \otimes H$ 使得 $m_H \circ (H \otimes \tau) \circ \gamma \circ \Delta = s \circ \varepsilon$.

一个带有反对极的 A-双代数胚称为 **A-Hopf 代数胚**.

一般地, 有如下定理 (见 [29]).

定理 4.7.14 设 H 是一个双代数, $(A, \cdot)$ 是一个左 H-模代数, (A, ρ^A) 是一个右 H-余模. 那么 $(A, \cdot, \rho^A) \in {}_H\mathbb{YD}^H$ 是一个辫子交换代数当且仅当 $(A\#H, s, t, \Delta, \varepsilon)$ 是一个 A-双代数胚, 结构映射为: 对任意 $a \in A, h \in H$,

$$s(a) = a\#1_H,$$

$$t(a) = \sum a_0 \# a_{(1)},$$

$$\Delta(a\#h) = a\#h_1 \otimes_A 1_A \#h_2,$$

$$\varepsilon(a\#h) = \varepsilon_H(h)a.$$

进一步, 如果 H 有反对极 S_H, 那么 $A\#H$ 是一个 A-Hopf 代数胚, 具有反对极:

$$\tau: A\#H \longrightarrow A\#H, \quad \tau(a\#h) = (S(h_2)S^2(a_{(1)})) \cdot a_0 \# S(h_1)S^2(a_{(2)})$$

对任意 $a \in A, h \in H$.

4.8 其他 Hopf 代数系

本节只简单地介绍一些我们工作中已经涉及的其他 Hopf 代数系, 目的为读者提供更多的信息资料.

4.8.1 H-伪代数与 H-伪余代数

除非特别注明, H 为余交换的 Hopf 代数.

定义 4.8.1 一个 H-**伪代数** (H-pseudoalgebra) 是一个左 H-模 A. 上面定义有运算 $\mu \in \text{Hom}_{H \otimes H}(A \otimes A, (H \otimes H) \otimes_H A)$, 叫做**伪积** (pseudoproduct).

我们时常将两个元素 $a, b \in A$ 的伪积 $\mu(a \otimes b) \in (H \otimes H) \otimes_H A$ 记为 $a * b$. 使用这个符号, 由定义, 伪积具有下面的性质.

(H-**双线性性**) 任取 $a, b \in A, f, g \in H$, 有

$$fa * b = ((f \otimes g) \otimes_H 1)(a * b).$$

4.8 其他 Hopf 代数系

即, 若设
$$a * b = \sum (f_i \otimes g_i) \otimes_H e_i,$$

则有
$$fa *_i gb = \sum_i (ff_i \otimes gg_i) \otimes_H e_i.$$

定义 4.8.2 称一个 H-伪代数 A 是结合的, 若任取 $a, b, c \in A$, 有

$$a * (b * c) = (a * b) * c.$$

称一个 H-伪代数 A 是交换的, 若任取 $a, b \in A$, 有

$$b * a = (\sigma \otimes_H \mathbf{id})(a \otimes b),$$

其中这里的 $\sigma: H \otimes H \longrightarrow H \otimes H$ 是换位映射 $\sigma(f \otimes g) = g \otimes f$.

下面给出一个 H-伪代数的例子:

例 4.8.3 设 H' 是 H 的子 Hopf 代数, 已知 A 是一个 H'-伪代数, 此时可以定义 Current H-伪代数 $\mathrm{Cru}_{H'}^H A$ 如下:

(i) 左 H-模 $H \otimes_{H'} A$;

(ii) 伪积定义如下: 任取 $a, b \in A$, 若 $a * b = \sum_i (f_i \otimes g_i) \otimes_{H'} e_i$, 则

$$(f \otimes_{H'} a) * (g \otimes_{H'} b) = ((f \otimes g) \otimes_H 1)(a * b) = \sum_i (ff_i \otimes gg_i) \otimes_H (1 \otimes_{H'} e_i).$$

易证此时 $\mathrm{Cru}_{H'}^H A$ 是一个 (结合的) H-伪代数当 A 是一个 (结合的)$H-$伪代数.

下面给出 H-伪代数的对偶定义, 若不特别说明, 下面出现的 H 均为结合的 Hopf 代数.

定义 4.8.4 一个 **H-伪余代数 (H-pseudocoalgebra)** C 满足

(i) C 是左 H-余模;

(ii) 存在左 $H \otimes H$-余模同态 $\Delta \in \mathrm{Hom}^{H \otimes H}(H \otimes H \otimes C, C \otimes C)$, 称作**伪余积 (pseudocoproduct)**.

为了简化书写, 引入下面的记号:

$$\Delta(f \otimes g \otimes c) = c_1^{f,g} \otimes c_2^{f,g}.$$

注意将 Δ 与 H 的余乘法 Δ_H 区分开.

定义 4.8.5 已知 H-伪余代数 C, 若有

$$\sigma \Delta = \Delta(\sigma \otimes \mathbf{id}),$$

即任取 $f,g \in H, c \in C$ 有

$$C_2^{f,g} \otimes C_1^{f,g} = C_1^{g,f} \otimes C_2^{g,f},$$

则称 C 为**余交换**H-伪余代数.

定义 4.8.6 已知 H-伪余代数 C, 若任取 $f,g,h \in H, c \in C$, 有

$$C_1^{f_2g_2,h}{}_1^{f_1,g_1} \otimes C_1^{f_2g_2,h}{}_2^{f_1,g_1} \otimes C_2^{f_2g_2,h} = C_1^{f,g_2h_2} \otimes C_2^{f,g_2h_2}{}_1^{g_1,h_1} \otimes C_2^{f,g_2h_2}{}_2^{g_1,h_1},$$

则称 C 是**余结合的**H-伪余代数.

例 4.8.7 下面给出一个 H-伪余代数的例子.

Hopf 代数 H 显然为本身的 H-余模, 定义伪余积: $\Delta(f \otimes g \otimes h) = f \otimes g$, 易证 H 为 H-伪余代数.

4.8.2 无穷小 Hopf 代数

这部分内容可参考文献 [3—6, 45—47, 80, 81, 113, 114, 259], 涉及组合问题的研究.

从文献可知, 交换非余交换的 Connes-Kreimer Hopf 代数用于处理量子场论 (Quantum Field Theory) 的重正规化 (renormalization) 问题, 而平面 (planar) 树的无穷小 Hopf 代数是这个 Connes-Kreimer Hopf 代数的推广. 同时, 非交换的情况也在文献 [91] 中研究.

定义 4.8.8 一个无穷小(infinitesimal)**双代数**(简写ε-**双代数**) 是一个三元数组 (A, m, Δ) 满足:

(i) (A, m) 是一个结合代数 (可能没有单位元);

(ii) (A, Δ) 是一个余结合余代数 (可能没有余单位元);

(iii) 对任意 $a, b \in A$, 有 $\Delta(ab) = (a \otimes 1)\Delta(b) + \Delta(a)(1 \otimes b)$, 换句话说, 余乘法 $\Delta: A \longrightarrow A \otimes A$ 是取值在 A-双模 $A \otimes A$ 上的 A 的导子 (derivation). 这等价于: 乘法 $m: A \otimes A \longrightarrow A$ 是余代数 A 的余导子 (coderivation).

定义 4.8.9 一个无穷小双代数称为是**无穷小 Hopf 代数**(简写ε-**Hopf 代数**) 如果存在一个映射 $S: A \longrightarrow A$ 满足, 对任意 $a \in A$

$$S(a_1)a_2 + S(a) + a = 0 = a_1 S(a_2) + a + S(a), \quad \forall a \in A.$$

在这个情况下, S 是唯一的, 被称为 A 的反对极.

一个 ε-双代数称为**分次的**, 如果存在一簇子空间 A_n 满足:

$$A = \bigoplus_{n=0}^{\infty} A_n, \quad m(A_i \otimes A_j) \subseteq A_{i+j}, \quad \Delta(A_n) \subseteq \bigoplus_{i+j=n-1} A_i \otimes A_j.$$

4.8 其他 Hopf 代数系

而且 A 是一个 ε-Hopf 代数具有

$$S = \sum_{n=1}^{\infty} (-1)^n m^{n-1} \Delta^{n-1}.$$

性质 4.8.10 (1) 设 A 是一个 ε-双代数. 如果 A 有单位 1, 那么 $\Delta(1) = 0$; 如果 A 既有单位又有余单位, 那么 $A = 0$.

(2) 设 A 是一个 ε-Hopf 代数具有反对极 S. 那么我们有:

(i) $S(1) = -1$, 如果存在单位元;

(ii) $S(ab) = -S(a)S(b)$, 对任意 $a, b \in A$;

(iii) $S(a_1) \otimes S(a_2) = -S(a)_1 \otimes S(a)_2$, 对任意 $a, b \in A$.

(3) 设 C 是一个余代数, A 是一个代数, 那么 $\mathrm{Hom}(C, A)$ 是一个幺半群具有循环卷积 (circular convolution product):

$$f \bullet g = f * g + f + g, \quad \text{即} \quad (f \bullet g)(c) = f(c_1)g(c_2) + f(c) + g(c).$$

单位元为 0 映射. 那么反对极 S 是恒等映射 $\mathrm{id}_A \in \mathrm{Hom}(A, A)$ 的逆.

(4) ε-Hopf 代数同态与一般的 Hopf 代数同态定义相同.

例 4.8.11 (1) 设 Q 是任意一个箭图 (有向图), 那么路代数 kQ 有一个 ε-双代数结构.

回顾 $kQ = \bigoplus_{n=0}^{\infty} kQ_n$, 这里 Q_n 是在 Q 中长度为 n 的路 γ 的集合:

$$\gamma : e_0 \xrightarrow{a_1} e_1 \xrightarrow{a_2} e_2 \xrightarrow{a_3} \cdots \to e_{n-1} \xrightarrow{a_n} e_n.$$

尤其, Q_0 是顶点集合, Q_1 是箭集合. 乘法是路的连接 (如果可以的话); 否则是零.
余乘法为: 对任意 $\gamma = a_1 a_2 \cdots a_n \in Q_n$,

$$\Delta(g) = e_0 \otimes a_2 a_3 \cdots a_n + a_1 \otimes a_3 \cdots a_n + \cdots + a_1 \cdots a_{n-1} \otimes e_n.$$

尤其, $\Delta(e) = 0$ 对任意 $e \in Q_0$, $\Delta(a) = s(a) \otimes t(a)$ 对任意 $a \in Q_1$.

路代数 kQ 是一个 ε-Hopf 代数. 反对极由性质 4.8.2 唯一确定为

$$S(e) = -e, \quad \forall e \in Q_0, \quad S(a) = \begin{cases} e - a, & s(a) = t(a) = e, \\ -a, & s(a) \neq t(a), \end{cases} \quad \forall a \in Q_1.$$

例如: 对于箭图.

我们有 ε-Hopf 代数是一个多项式代数 $k[x]$. 我们有 $k[x] \otimes k[x] \cong k[x, y]$, 余乘法是一个可除差分算子 (divided difference operator):

$$\Delta(f(x)) = \frac{f(x) - f(y)}{x - y}.$$

反对极 S 为
$$S(f(x)) = -f(x-1).$$

(2) 设 $A = k\langle x_1, x_2, x_3, \cdots \rangle$ 是一个自由代数, 那么 A 具有 ε-Hopf 代数结构:
$$\Delta(x_n) = \sum_{i=0}^{n-1} x_i \otimes x_{n-1-i} = 1 \otimes x_{n-1} + x_1 \otimes x_{n-2} + \cdots + x_{n-1} \otimes 1.$$
$$S(x_n) = \sum_{k=1}^{n+1} (-1)^k \sum_{n_1+\cdots+n_k=n+1; n_i>0} x_{n_1-1} x_{n_2-1} \cdots x_{n_k-1},$$

这里 $x_0 := 1$.

4.8.3 乘子 Hopf 代数

这方面研究作者可以参考文献 [55, 59, 60, 70, 104, 191, 195, 196, 204, 256, 258], 已经有许多成果发表, 也是弱乘子 Hopf 代数今后研究的主流方向.

4.8.4 无穷小乘子 Hopf 代数

主要起点工作见文献 [14], 也是今后研究的问题之一, 无穷小乘子 Hopf 代数与李双代密切相关.

第 5 章 弱乘子 Hopf 代数

本章主要介绍弱乘子 Hopf 代数的最新研究成果, 包括积分理论、对偶理论等, 也介绍了乘子 Hopf 代数胚的基本概念. 这些概念与理论已经形成了国际该领域研究的热点问题之一, 见参考文献 [22—24, 199—202, 280].

5.1 定义与例子

首先我们给出一些预备知识.

在本章中, 所有的代数都是假定在复数域 $\mathbb{C}$ 上. 设 A 是一个代数不一定含有单位, 但是 A 带有非退化的乘法, 即对任意的 $a \in A$ 且 $ab = 0$ 对任意的 b 或者 $ba = 0$ 对任意的 $b \in A$, 一定可以得到 $a = 0$. 对于 A, 我们可以考虑它的乘子代数 $M(A)$. A 可以看作是 $M(A)$ 中的一个稠密的双边理想. 如果 A 是非退化的, 那么 $A \otimes A$ 也是. 并且有如下自然嵌入:

$$A \otimes A \subseteq M(A) \otimes M(A) \subseteq M(A \otimes A).$$

在本章中, 我们还要假定 A 是幂等的. A 称为幂等代数如果 $A^2 = A$.

设 A 是一个带有非退化乘法的代数. 对任意的 $a, b, c \in A$, 同态 $\Delta: A \longrightarrow M(A \otimes A)$ 称为 A 上的**余乘**(coproduct) 如果满足

(i) $\Delta(a)(1 \otimes b) \in A \otimes A$ 和 $(a \otimes 1)\Delta(b) \in A \otimes A$;

(ii) Δ 是余结合的, 即满足

$$(c \otimes 1 \otimes 1)(\Delta \otimes \iota)(\Delta(a)(1 \otimes b)) = (\iota \otimes \Delta)((c \otimes 1)\Delta(a))(1 \otimes 1 \otimes b).$$

余乘 Δ 称为**完全**(full) 的如果满足下列条件:

$$\Delta(A)(1 \otimes A) \subseteq V \otimes A \quad \text{和} \quad (A \otimes 1)\Delta(A) \subseteq A \otimes W$$

的最小子空间 V, W 都是 A. 余乘 Δ 称为**正则**(regular) 的如果对任意的 $a, b \in A$, 有 $\Delta(a)(b \otimes 1) \in A \otimes A$, $(1 \otimes a)\Delta(b) \in A \otimes A$.

给定一个余乘 Δ, 对任意的 $a, b \in A$, 可以定义典范映射 T_1, T_2 为

$$T_1(a \otimes b) = \Delta(a)(1 \otimes b), \quad T_2(a \otimes b) = (a \otimes 1)\Delta(b).$$

如果 Δ 是正则的, 则可定义 T_3, T_4 为

$$T_3(a \otimes b) = (1 \otimes b)\Delta(a), \quad T_4(a \otimes b) = \Delta(b)(a \otimes 1).$$

线性映射 $\varepsilon: A \to \mathbb{C}$ 称为 A 上的**余单位**如果满足

$$(\varepsilon \otimes \mathbf{id})(\Delta(a)(1 \otimes b)) = ab \quad \text{和} \quad (\mathbf{id} \otimes \varepsilon)((a \otimes 1)\Delta(b)) = ab,$$

对任意的 $a, b \in A$.

对任意向量空间 A, 记 $\mathbf{fl}: A \otimes A \longrightarrow A \otimes A, a \otimes b \mapsto b \otimes a$. 对余乘 $\Delta: A \longrightarrow M(A \otimes A)$, 定义 $\Delta^{\text{cop}}: A \longrightarrow M(A \otimes A)$ 为

$$\Delta^{\text{cop}}(a)(b \otimes c) := \mathbf{fl}(\Delta(a)(c \otimes b)) \quad \text{和} \quad (b \otimes c)\Delta^{\text{cop}}(a) := \mathbf{fl}((c \otimes b)\Delta(a)),$$

定义 $\Delta_{13}: A \longrightarrow M(A \otimes A \otimes A)$ 为

$$\Delta_{13}(a)(b \otimes c \otimes d) := (\mathbf{id} \otimes \mathbf{fl})(\Delta(a)(b \otimes d) \otimes c),$$

$$(b \otimes c \otimes d)\Delta_{13}(a) := (\mathbf{id} \otimes \mathbf{fl})((b \otimes d)\Delta(a) \otimes c)$$

对任意的 $a, b, c, d \in A$.

在本章中我们将采用 Sweddler 记号.

定义 5.1.1 设 A 是一个非退化的幂等代数并且带有一个满的余乘 Δ 和一个余单位 ε, 如果 A 满足以下条件则 A 被称为**弱乘子 Hopf 代数**(weak multiplier Hopf algebra):

(i) 存在一个幂等元素 $E \in M(A \otimes A)$ 使得

$$E(A \otimes A) = T_1(A \otimes A),$$

$$(A \otimes A)E = T_2(A \otimes A);$$

(ii) 元素 E 满足

$$(\mathbf{id} \otimes \Delta)E = (E \otimes 1)(1 \otimes E) = (1 \otimes E)(E \otimes 1);$$

(iii) 典范映射 T_1 和 T_2 的核满足

$$\text{Ker}(T_1) = (1 - G_1)(A \otimes A),$$

$$\text{Ker}(T_2) = (1 - G_2)(A \otimes A),$$

其中 G_1 和 G_2 是从 $A \otimes A$ 到 $A \otimes A$ 的线性映射, 定义为

$$(G_1 \otimes \mathbf{id})(\Delta_{13}(a)(1 \otimes b \otimes c)) = \Delta_{13}(a)(1 \otimes E)(1 \otimes b \otimes c),$$

5.1 定义与例子

$$(\mathrm{id} \otimes G_2)((a \otimes b \otimes 1)\Delta_{13}(c)) = (a \otimes b \otimes 1)(E \otimes 1)\Delta_{13}(c)$$

对任意的 $a, b, c \in A$.

定义中的 E 被称为**典范幂等元**(canonical idempotent), 并且我们有 $E = \Delta(1)$. 一个弱乘子 Hopf 代数 (A, Δ) 称为是**正则**的如果 Δ 是正则的并且 $(A, \Delta^{\mathrm{op}})$ 也是一个弱乘子 Hopf 代数, 或者要求 $(A^{\mathrm{op}}, \Delta)$ 是一个弱乘子 Hopf 代数.

如果 A 同时是 $*$-代数并且 Δ 是 $*$-同态, 那么 A 称为**弱乘子 Hopf $*$-代数**(weak multiplier Hopf $*$-algebra), 此时正则性自然成立.

接下来我们将给出两个例子.

例 5.1.2 设 G 是一个群胚, 考虑 G 上的所有有限支撑复泛函组成的 $*$-代数 $K(G)$. $*$-运算定义为 $f^*(p) = \overline{f(p)}, f \in K(G), p \in G$. 记这个 $*$-代数为 A. A 是单位代数当且仅当 G 是有限的, 并且 A 是非退化的幂等代数. 那么接下来由 G 中的乘法我们可以定义 A 上的余乘 Δ 为

$$\Delta(f)(p, q) = \begin{cases} f(pq), & pq \text{ 有定义}, \\ 0, & \text{其他}, \end{cases}$$

Δ 是一个从 A 到 $M(A \otimes A)$ 的 $*$-同态.

令 $f, g \in A$, 考虑泛函 $\Delta(f)(1 \otimes g)$. 有

$$\Delta(f)(1 \otimes g)(p, q) = \begin{cases} f(pq)g(q), & pq \text{ 有定义}, \\ 0, & \text{其他}. \end{cases}$$

由于 $g \in A$, 有 q 属于有限集, 否则上述泛函为 0, 同时 pq 一定属于有限集. 因为 $p = (pq)q^{-1}$, 所以 $f(pq)g(q)$ 为 0 如果 p 不属于某个有限集. 综上可得 $\Delta(f)(1 \otimes g) \in K(G \times G)$. 类似地我们也可以得到 $(f \otimes 1)\Delta(g) \in K(G \times G)$, 即定义 5.1.1 的条件 (i) 满足.

由 G 中乘法的结合性可得到 Δ 的余结合性. 又因为 A 是交换代数, 所以余乘也是正则的.

余单位由 $\varepsilon(f) = \sum f(e)$ 给出, 这里求和是对 G 中所有单位求和. 接下来证明 Δ 是满的. 对任意 $p \in G$, 设 e 是 p 的靶. 令

$$f = \delta_p = \begin{cases} 1, & p, \\ 0, & \text{其他}, \end{cases}$$

那么有 $\Delta(f)(e, \cdot) = f$, 这就表明 Δ 的右腿是 A. 类似地, 考虑一个元素的源, 我们可以证明 Δ 的左腿是 A.

E 定义为 $G \times G$ 上的一个泛函, 当 pq 有定义时在 (p, q) 处值为 1, 否则为 0. 容易证明 E 满足定义中的条件. G_1 和 G_2 由 $M(A \otimes A)$ 中的幂等元 F_1, F_2 给出.

F_1, F_2 都是 $G \times G$ 上的泛函. F_1 定义为

$$F_1(p,q) = \begin{cases} 1, & s(p) = s(q), \\ 0, & \text{其他}. \end{cases}$$

F_2 定义为

$$F_2(p,q) = \begin{cases} 1, & t(p) = t(q), \\ 0, & \text{其他}. \end{cases}$$

从而有 $(E_{13}(F_1 \otimes 1))(p,q,v) = 1$ 当且仅当 $s(p) = t(v), s(p) = s(q)$. 另一方面根据 G_1 的定义有 $(E_{13}(F_1 \otimes 1))(p,q,v) = 1$ 当且仅当 $s(p) = t(v), s(q) = t(v)$. 这就表明 G_1 由 F_1 给出. 类似地, G_2 由 F_2 给出. 不难证明 G_1, G_2 满足定义 5.1.1 的条件 (iii).

由于 A 是交换的, 最终得到了一个正则的弱乘子 Hopf 代数.

接下来考虑上述例子的对偶.

例 5.1.3 设 G 是一个群胚, 考虑 G 上所有有限支撑复泛函构成的代数 $\mathbb{C}G$, 乘法为卷积. 记这个代数为 A. 令 $p \mapsto \lambda_p$ 为 G 到 $\mathbb{C}G$ 的典范嵌入, 对任意的 p, q, 当 pq 有定义时有 $\lambda_p \lambda_q = \lambda_{pq}$, 其他则乘积为 0. 如果定义 $\lambda_p^* = \lambda_{p^{-1}}$, 则 A 是 $*$-代数. A 是单位代数当且仅当 G 有有限个单位.

A 上的余乘定义为 $\Delta(\lambda_p) = \lambda_p \otimes \lambda_p$, 对任意的 $p \in G$. 余单位定义为 $\varepsilon(\lambda_p) = 1$, 对任意的 p. 典范幂等元 E 定义为 $\sum \lambda_e \otimes \lambda_e$, 这里是对 G 中所有单位求和. 我们可以证明 E 是满足下列条件的最小幂等元:

$$E(\lambda_p \otimes \lambda_p) = \lambda_p \otimes \lambda_p \quad \text{和} \quad (\lambda_p \otimes \lambda_p)E = \lambda_p \otimes \lambda_p$$

对任意的 $p \in G$. 显然

$$(\Delta \otimes \mathbf{id})(E) = (\mathbf{id} \otimes \Delta)(E) = \sum_e \lambda_e \otimes \lambda_e \otimes \lambda_e.$$

同时还可以得到

$$E \otimes 1 = \sum_{e,f} \lambda_e \otimes \lambda_e \otimes \lambda_f \quad \text{和} \quad 1 \otimes E = \sum_{e,f} \lambda_f \otimes \lambda_e \otimes \lambda_e,$$

这里是对单位求和. 从上可以得到 $E \otimes 1$ 和 $1 \otimes E$ 交换并且乘积为 $\sum_e \lambda_e \otimes \lambda_e \otimes \lambda_e$.

可以证明 G_1 和 G_2 都由 F_1, F_2 给出, 其中 $F_1 = F_2 = \sum_e \lambda_e \otimes \lambda_e$. 不难证明 G_1, G_2 满足定义.

综上, 结合 A 为余交换的, 可得 A 是一个正则的弱乘子 Hopf 代数.

对任意的一个弱乘子 Hopf 代数 A, 均存在一个反代数同态和反余代数同态 $S: A \to M(A)$ 满足

$$\sum S(a_{(1)})a_{(2)}S(a_{(3)}) = S(a),$$

5.1 定义与例子

$$\sum a_{(1)} S(a_{(2)}) a_{(3)} = a$$

对任意的 $a \in A$. S 称为**对极**.

给定一个正则的弱乘子 Hopf 代数 $(A, \Delta, \varepsilon, E, S)$, 对任意的 $a \in A$, 可以定义如下四种从 A 到 $M(A)$ 的映射:

$$\begin{cases} \varepsilon_s(a) = (1 \otimes \varepsilon)((1 \otimes a)E) = \sum S(a_{(1)}) a_{(2)}, \\ \varepsilon_t(a) = (\varepsilon \otimes 1)(E(a \otimes 1)) = \sum a_{(1)} S(a_{(2)}), \\ \varepsilon'_s(a) = (1 \otimes \varepsilon)(E(1 \otimes a)) = \sum a_{(2)} S^{-1}(a_{(1)}), \\ \varepsilon'_t(a) = (\varepsilon \otimes 1)((a \otimes 1)E) = \sum S^{-1}(a_{(2)}) a_{(1)}. \end{cases}$$

$\varepsilon_s(A)$ 称为**源代数**, $\varepsilon_t(A)$ 称为**靶代数**(target algebra). 当 A 是正则的情况下, 我们可以考虑源代数和靶代数的乘子代数, 记它们的乘子代数分别为 $M(\varepsilon_s(A)), M(\varepsilon_t(A))$, 定义如下

$$M(\varepsilon_s(A)) = \{y \in M(A) \mid \Delta(ay)(1 \otimes b) = \Delta(a)(1 \otimes yb) \text{ 对任意的 } a, b \in A\},$$
$$M(\varepsilon_t(A)) = \{x \in M(A) \mid (c \otimes 1)\Delta(xa) = (cx \otimes 1)\Delta(a) \text{ 对任意的 } a, c \in A\}.$$

接下来给出弱乘子 Hopf 代数的另一个刻画, 证明见 [239].

定理 5.1.4 设 (A, Δ, ε) 是一个非退化代数带有一个完全的余乘和余单位. 如果满足下列条件:

(i) 存在一个线性映射 $S: A \to M(A)$ 使得

$$R_1(a \otimes b) = \sum a_{(1)} \otimes S(a_{(2)}) b,$$
$$R_2(a \otimes b) = \sum a S(b_{(1)}) \otimes b_{(2)}$$

都是属于 $A \otimes A$, 并且对任意的 $a \in A$ 有

$$\sum S(a_{(1)}) a_{(2)} S(a_{(3)}) = S(a) \quad \text{和} \quad \sum a_{(1)} S(a_{(2)}) a_{(3)} = a$$

成立.

(ii) 存在一个元素 $E \in M(A \otimes A)$ 使得

$$(T_1 R_1)(a \otimes b) = E(a \otimes b), \qquad (T_2 R_2)(a \otimes b) = (a \otimes b)E$$

对任意的 $a, b \in A$. E 还要满足

$$(\mathbf{id} \otimes \Delta)(E) = (\Delta \otimes \mathbf{id})(E) = (E \otimes 1)(1 \otimes E) = (1 \otimes E)(E \otimes 1).$$

那么 A 就是弱乘子 Hopf 代数, 并且 S 是 A 的对极.

命题 5.1.5 如果 (A, Δ) 是正则的弱乘子 Hopf 代数, 那么 A 有局部单位.

证明 对任意 $a \in A$, 假设 ω 是 A 上的一个线性泛函, 并且对任意的 $b \in A$ 有 $\omega(ab) = 0$. 接下来我们会证明 $\omega(a) = 0$, 这样就得到 $a \in aA$. 类似地, 我们可以得到 $a \in Aa$, 这样就证明了 A 有局部单位.

假设 ω 在 aA 上为 0. 由 [200] 中的命题 3.9 可得 $\omega(ay) = 0$, 对任意的 $y \in \varepsilon_s(A)$. 根据 [200] 中引理 3.2 有

$$(\omega \otimes \mathbf{id})(a \otimes 1)E(y \otimes 1) = 0.$$

两端同时作用 S, 利用 $F_1 = (\mathbf{id} \otimes S)E$ 可得

$$(\omega \otimes \mathbf{id})(a \otimes 1)F_1(y \otimes 1) = 0.$$

又因为

$$(a \otimes 1)F_1(1 \otimes b) = R_1 T_1(a \otimes b) = \sum a_{(1)} \otimes S(a_{(2)})a_{(3)}b$$

对任意的 $b \in A$. 从而有

$$\sum \omega(a_{(1)}y) S(a_{(2)})a_{(3)} = 0$$

对任意的 $p, q \in A$ 有

$$(pq \otimes 1)F_1 = (p \otimes 1)(q \otimes 1)F_1 - \sum pq_{(1)} \otimes S(q_{(2)})q_{(3)}.$$

又因为 $\sum a_{(1)} \otimes S(a_{(2)})a_{(3)} \in A \otimes \varepsilon_s(A)$, 应用赋值映射可得

$$\sum \omega(a_{(1)}S(a_{(2)})a_{(3)}) = 0,$$

即 $\omega(a) = 0$.

命题 5.1.6 A 是正则的当且仅当 S 是双射.

证明 见 [199] 命题 4.11.

命题 5.1.7 设 (A, Δ) 是弱 Hopf 代数, 那么 (A, Δ) 是一个弱乘子 Hopf 代数.

证明 根据假设, A 是一个带有单位的代数, 自然也是非退化和幂等的. 余乘也满足我们的定义. 根据假设, 存在一个余单位, 可以得到 Δ 是完全的. 在这种情况下, 不难证明 Δ 是也正则的.

根据假设, 存在一个对极 S. 我们将证明 S 满足定理 5.1.4 的条件 (i). 首先 S 是从 A 到 A 的一个线性映射, 对应地

$$R_1(a \otimes b) = \sum a_{(1)} \otimes S(a_{(2)})b,$$

5.1 定义与例子

$$R_2(a\otimes b)=\sum aS(b_{(1)})\otimes b_{(2)}$$

都是属于 $A\otimes A$. 条件 (i) 中的第二个等式是弱 Hopf 代数定义的一部分. 接下来我们证明第一个等式. 对任意的 $a\in A$, 有

$$\sum a_{(1)}S(a_{(2)})a_{(3)}=\sum (\varepsilon\otimes \mathbf{id})(\Delta(1)(a_{(1)}\otimes 1))a_{(2)}$$
$$=\sum (\varepsilon\otimes \mathbf{id})(\Delta(1)(a_{(1)}\otimes a_{(2)}))$$
$$=(\varepsilon\otimes \mathbf{id})\Delta(a)=a.$$

令 $E=\Delta(1)$. 根据定义, 定理 2.9 条件 (ii) 的第二部分自然满足. 接下来我们证明定理 2.9 条件 (ii) 的第一部分. 对任意的 $a\in A$, 有

$$T_{(1)}R_{(1)}(a\otimes 1)=\sum a_{(1)}\otimes a_{(2)}S(a_{(3)})$$
$$=\sum a_{(1)}\otimes (\varepsilon\otimes \mathbf{id})(\Delta(1)(a_{(2)}\otimes 1))$$
$$=(\mathbf{id}\otimes \varepsilon\otimes \mathbf{id})((1\otimes \Delta(1))(\Delta(a)\otimes 1))$$
$$=(\mathbf{id}\otimes \varepsilon\otimes \mathbf{id})((1\otimes \Delta(1))(\Delta(1)\otimes 1)(\Delta(a)\otimes 1))$$
$$=(\mathbf{id}\otimes \varepsilon\otimes \mathbf{id})((\Delta\otimes \mathbf{id})(\Delta(1)))(\Delta(a)\otimes 1)$$
$$=(\mathbf{id}\otimes \varepsilon\otimes \mathbf{id})((\Delta\otimes \mathbf{id})(\Delta(1)(a\otimes 1)))$$
$$=\Delta(1)(a\otimes 1),$$

这就证明了 $T_1R_1(a\otimes b)=E(a\otimes b)$. 类似地, 我们可以得到 $T_2R_2(a\otimes b)=(a\otimes b)E$.

命题 5.1.8 设 (A,Δ) 是一个正则的弱乘子 Hopf 代数, 若 A 有单位, 则 A 是弱 Hopf 代数.

证明 因为 A 有单位, 则 $M(A)=A, M(A\otimes A)=A\otimes A$. 显然有 $E=\Delta(1)$. 此时对极也是从 A 到 A 的一个映射. 由于 $E(a\otimes 1)=\sum a_{(1)}\otimes a_{(2)}S(a_{(3)})$, 作用 $(\varepsilon\otimes \mathbf{id})$ 可得

$$(\varepsilon\otimes \mathbf{id})(\Delta(1)(a\otimes 1))=\sum a_{(1)}S(a_{(2)})$$

对任意的 $a\in A$. 类似地, 有 $(1\otimes a)E=\sum S(a_{(1)})a_{(2)}\otimes a_{(3)}$, 作用 $(\mathbf{id}\otimes \varepsilon)$ 可得

$$(\mathbf{id}\otimes \varepsilon)((1\otimes a)\Delta(1))=\sum S(a_{(1)})a_{(2)}$$

对任意的 $a\in A$. 又 $\sum S(a_{(1)}a_{(2)}S(a_{(3)}))=S(a)$, 可以得到 S 满足所有条件.

接下来证明 ε 满足定义所有条件, 即证明

$$\varepsilon(abc)=\sum \varepsilon(ab_{(2)})\varepsilon(b_{(1)}c),$$
$$\varepsilon(abc)=\sum \varepsilon(ab_{(1)})\varepsilon(b_{(2)}c)$$

对任意的 $a, b, c \in A$. 首先我们证明第一个等式. 对任意的 $a, b, c \in A$, 有

$$(1 \otimes a)\Delta(b)(c \otimes 1) = \sum (1 \otimes a)\Delta(bc_{(1)}(1 \otimes S(c_{(2)})))$$

对上式两端同时作用 $(\varepsilon \otimes \varepsilon)$ 可得

$$(\varepsilon \otimes \varepsilon)((1 \otimes a)\Delta(b)(c \otimes 1)) = \sum \varepsilon(abc_{(1)}S(c_{(2)})).$$

令 $a = 1$, 可得

$$\varepsilon(bc) = (\varepsilon \otimes \varepsilon)\Delta(b)(c \otimes 1) = \sum \varepsilon(bc_{(1)}S(c_{(2)})).$$

利用这个等式并且在之前的等式中将 b 替换为 ab, 可得

$$(\varepsilon \otimes \varepsilon)((1 \otimes b)\Delta(a)(c \otimes 1)) = \varepsilon(abc)$$

对任意的 $a, b, c \in A$. 这就证明了第一个等式.

因为 A 是正则的, 可以对 $(A, \Delta^{\mathrm{cop}})$ 重复上述证明, 所以可得到

$$(\varepsilon \otimes \varepsilon)((a \otimes 1)\Delta(b)(1 \otimes c)) = \varepsilon(abc).$$

命题 5.1.9 设 $(A \Delta)$ 是一个有限维弱 Hopf 代数, 那么它是一个正则的弱乘子 Hopf 代数. 反过来, 如果 $(A \Delta)$ 是一个有限维的正则的弱乘子 Hopf 代数, 那么 A 是一个有限维的弱 Hopf 代数.

证明 由命题 5.1.7 可知, 如果 A 是一个有限维弱 Hopf 代数, 那么 A 就是一个弱乘子 Hopf 代数. 因为有限维的弱 Hopf 代数的对极是双射, 由命题 5.1.6 知 A 是正则的.

反过来, 由于 A 是正则的, 所有 A 有局部单位, 又 A 是有限维的, 所有 A 有单位. 根据命题 5.1.8 的结果可得 A 一定是弱 Hopf 代数.

5.2 弱乘子双代数

在本节中, 我们将介绍弱乘子双代数[24]. 并且主要讨论弱乘子双代数、弱 Hopf 代数与弱乘子 Hopf 代数之间的关系.

定义 5.2.1 设 A 是一个幂等的非退化的代数, 称 A 为**弱乘子双代数**(weak multiplier bialgebra), 如果存在一个完全的余乘 Δ、幂等元 $E \in M(A \otimes A)$ 和余单位 ε 满足下列条件:

(i)
$$\langle \Delta(a)(b \otimes b') \mid a, b, b' \in A \rangle = \langle E(b \otimes b') \mid b, b' \in A \rangle,$$

5.2 弱乘子双代数

$$\langle (b \otimes b')\Delta(a) \mid a, b, b' \in A \rangle = \langle (b \otimes b')E \mid b, b' \in A \rangle.$$

(ii) E 满足

$$(E \otimes 1)(1 \otimes E) = E^{(3)} = (1 \otimes E)(E \otimes 1),$$

这里 $E^{(3)} := (\mathbf{id} \otimes \Delta)E = (\Delta \otimes \mathbf{id})E.$

(iii) $\forall a, b, c \in A$, 有

$$(\varepsilon \otimes \mathbf{id})((1 \otimes a)E(b \otimes c)) = (\varepsilon \otimes \mathbf{id})(\Delta(a)(b \otimes c)),$$

$$(\varepsilon \otimes \mathbf{id})((a \otimes b)E(1 \otimes c)) = (\varepsilon \otimes \mathbf{id})((a \otimes b)\Delta(c)).$$

弱乘子双代数 A 称为是正则的如果下列元素

$$T_3(a \otimes b) := (1 \otimes b)\Delta(a) \quad \text{和} \quad T_4(a \otimes b) := \Delta(b)(a \otimes 1)$$

都属于 $A \otimes A$, 对任意的 $a, b \in A$.

弱乘子双代数两类重要的例子就是正则的弱乘子 Hopf 代数和弱双代数, 见下面两个定理, 证明见文献 [24].

定理 5.2.2 如果 $(A, \Delta, \varepsilon, E)$ 是正则的弱乘子 Hopf 代数, 那么在同样的结构下 A 满足弱乘子双代数的定义.

定理 5.2.3 对任意一个单位代数 A, 下面叙述等价:
(i) A 有一个弱双代数结构;
(ii) A 有一个弱乘子双代数结构.

接下来我们定义弱乘子双代数上的反对极, 并得到本节的主要定理. 设 A 是一个正则的弱乘子双代数, 由 [200] 中命题 4.3 我们可以定义 $F: A \otimes A \to M(A \otimes A)$ 为

$$a \otimes bc \mapsto ((\varepsilon'_s \otimes \mathbf{id})T_4^{\mathrm{op}}(c \otimes b))(a \otimes 1),$$
$$ab \otimes c \mapsto (1 \otimes c)((\mathbf{id} \otimes \varepsilon_s)T_2(a \otimes b)).$$

由 A 的幂等性, 可得 $(a \otimes 1)F \in A \otimes M(A)$. 从而我们可以定义

$$G_1: A \otimes A \to A \otimes A, \quad a \otimes b \mapsto (a \otimes 1)F(1 \otimes b).$$

类似地, 可以定义

$$G_2: ab \otimes c \mapsto ((\mathbf{id} \otimes \varepsilon'_t)T_3^{\mathrm{op}}(b \otimes a))(1 \otimes c).$$

定理 5.2.4 设 A 是一个正则的弱乘子双代数, 在下列两个论断中存在一个一一映射.

(1) 对任意的 $i \in \{1,2\}$, 存在线性映射 $R_i : A \otimes A \to A \otimes A$ 使得 $R_iT_i = G_i, T_iR_i = E_i, R_iT_iR_i = R_i$.

(2) 存在一个线性映射 $S : A \to M(A)$ 满足

(i) $T_1[((\mathbf{id} \otimes S)T_2(a \otimes b))(1 \otimes c)] = \Delta(a)(b \otimes c)$;

(ii) $T_2[(a \otimes 1)(S \otimes \mathbf{id})T_1(b \otimes c)] = (a \otimes b)\Delta(c)$;

(iii) $\mu(S \otimes \mathbf{id})[E(a \otimes 1)] = S(a)$(等价地,$\mu(\mathbf{id} \otimes S)[(1 \otimes a)E] = S(a)$) 对任意的 $a,b,c \in A$.

证明 (i) 推出 (ii). 由 [24] 中推论 6.7 存在一个线性映射 $(\lambda_{R_1}, \rho_{R_2}) =: A \to M(A)$. 利用 $T_1R_1 = E_1$, 可得

$$\begin{aligned}T_1[((\mathbf{id} \otimes S)T_2(a \otimes b))(1 \otimes c)] &= T_1[(a \otimes 1)R_1(b \otimes c)] \\ &= \Delta(a)T_1R_1(b \otimes c) \\ &= \Delta(a)E(b \otimes c) \\ &= \Delta(a)(b \otimes c)\end{aligned}$$

对任意的 $a,b,c \in A$. 类似地, 利用 $T_2R_2 = E_2$, 可以推出 (2). 利用 $R_1E_1 = R_1$, 对任意的 $a,b \in A$, 可得

$$\mu(S \otimes \mathbf{id})[E(a \otimes b)] = \lambda_{R_1E_1}(a)b = \lambda_{R_1}(a)b = S(a)b.$$

(ii) 推出 (i). 由定义 5.2.1 的 (i) 可得

$$\mathrm{Im}(T_1) \subseteq \langle E(a \otimes b)|a,b \in A \rangle = \langle \Delta(a)(b \otimes c)|a,b,c \in A \rangle.$$

反过来, 由本定理中的 (1) 有 $\langle \Delta(a)(b \otimes c)|a,b,c \in A \rangle \subseteq \mathrm{Im}(T_1)$, 所以有 $\mathrm{Im}(T_1) = \mathrm{Im}(E_1)$. 由 [200] 中的命题 6.3 的 (3), $T_1G_1 = T_1$, 有 $\mathrm{Ker}(G_1) \subseteq \mathrm{Ker}(T_1)$. 在 (2) 式两端作用 $(\mathbf{id} \otimes \varepsilon)$ 并结合余单位性质和 A 是幂等的, 可以得到

$$\mu(S \otimes \mathbf{id})T_1 = \mu(\varepsilon_s \otimes \mathbf{id}).$$

假设存在 $b,c \in A$ 使得 $T_1(b \otimes c) = 0$. 那么对任意的 $a \in A$ 有

$$\begin{aligned}0 &= (\mathbf{id} \otimes \mu)(\mathbf{id} \otimes S \otimes \mathbf{id})(T_2 \otimes \mathbf{id})(\mathbf{id} \otimes T_1)(a \otimes b \otimes c) \\ &= (\mathbf{id} \otimes \mu)(\mathbf{id} \otimes S \otimes \mathbf{id})(\mathbf{id} \otimes T_1)(T_2 \otimes \mathbf{id})(a \otimes b \otimes c) \\ &= (\mathbf{id} \otimes \mu)(\mathbf{id} \otimes \varepsilon_s \otimes \mathbf{id})(T_2 \otimes \mathbf{id})(a \otimes b \otimes c) \\ &= G_1(ab \otimes c) = (a \otimes 1)G_1(b \otimes c).\end{aligned}$$

因为 $A \otimes A$ 是非退化的, 可得 $G_1(b \otimes c) = 0$. 从而 $\mathrm{Ker}(G_1) = \mathrm{Ker}(T_1)$.

根据 [201] 中的命题 2.3, 上述证明可推出存在线性映射 $R_1 : T_1(a \otimes b) \mapsto G_1(a \otimes b)$ 满足我们的条件. 类似地, 可以构造出 R_2.

接下来证明上述对应是一个双射. 从 R_1 的表达式来看, 它并不依赖于 S 的具体作用 (只依赖于 S 的存在性). 因此, 从 (i) 出发, 我们得到 $R_1T_1 = G_1$, R_1 必须按照上述表达式来定义. 又因为 $R_1 - R_1E_1$, R_1 作用在 $\mathrm{Ker}(E_1)$ 上一定为 0. 对于 R_2 我们可以得到类似的结论. 反过来, 如果从 (ii) 中的 S 出发, 重复上述构造过程 $S \mapsto (R_1, R_2) \mapsto (\lambda_{R_1}, \rho_{R_2})$, 我们可以得到映射 $\lambda_{R_1} : A \to \mathrm{End}_A(A)$ 将 $a \in A$ 映为
$$b \mapsto (\varepsilon \otimes \mathbf{id})R_1(a \otimes b) = (\varepsilon \otimes \mathbf{id})G_1R_1(a \otimes b) = \mu(S \otimes \mathbf{id})T_1R_1(a \otimes b) = \mu(S \otimes \mathbf{id})[E(a \otimes b)].$$

对任意的 $a, b \in A$, 元素 $\lambda_{R_1}(a)b$ 等于 $S(a)b$ 当且仅当 (3) 成立. 类似地, $a\rho_{R_2}(b)$ 等于 $aS(b)$ 当且仅当 (3) 中第二个等式成立, 从而证明了 (3) 中两个等式是等价的.

定理 5.2.4 也表明, 如果反对极存在, 一定是唯一的. 结合定理 5.2.2 和定理 5.2.4, 我们可以得出 5.1 节中的正则的弱乘子 Hopf 代数可以看成是一个正则的弱乘子双代数加上一个反对极. 另一方面, 如果一个正则的弱乘子双代数上存在一个反对极, 那么它一定是一个弱乘子 Hopf 代数, 不一定是正则的. 也就是说, 带有反对极的正则的弱乘子双代数是一类介于任意的弱乘子 Hopf 代数和正则的弱乘子 Hopf 代数之间的代数.

5.3　Pontryagin 对偶

设 A 是一个代数量子群胚, 本节将构造其对偶 $\hat{A}$ 并证明对偶上面仍然有一个代数量子群胚结构. 此外通过考虑 $\hat{A}$ 的对偶, 发现与 A 同构, 这就是 Pontryagin 对偶定理在弱乘子 Hopf 代数情况下的表现.

首先我们介绍弱乘子 Hopf 代数的积分理论.

设 A 是正则的弱乘子 Hopf 代数, 对任意的 $a \in A$ 以及 A 上的泛函 ω. 我们可以定义乘子 $x \in M(A)$ 如下
$$xb = (\mathbf{id} \otimes \omega)(\Delta(a)(b \otimes 1)) \quad \text{和} \quad bx = (\mathbf{id} \otimes \omega)(b \otimes 1)\Delta(a)$$
对任意的 $b \in A$. 乘子 x 记为 $(\mathbf{id} \otimes \omega)\Delta(a)$. 类似地, 我们可以定义 $(\omega \otimes \mathbf{id})\Delta(a) \in M(A)$ 对任意的 $a \in A$ 和任意的泛函 ω.

在本节中记 $A_t = M(\varepsilon_t(A)), A_s = M(\varepsilon_s(A))$.

定义 5.3.1　线性泛函 $\varphi : A \to \mathbb{C}$ 称为**左不变**的如果 $(\mathbf{id} \otimes \varphi)\Delta(a) \in A_t$ 对任意的 $a \in A$. 类似地, 线性泛函 ψ 称为**右不变**的如果 $(\psi \otimes \mathbf{id})\Delta(a) \in A_s$ 对任意的 $a \in A$. 一个非零的左不变泛函称为 A 上的**左积分**, 一个非零的右不变泛函称为 A 上的**右积分**.

注记 5.3.2 (i) 如果 A 是一个乘子 Hopf 代数, 此时 A_t 和 A_s 恰好是 1 的倍数, 乘子 Hopf 代数上的积分的定义正好符合此时的情况.

(ii) 由于对极 S 将 A_t 映到 A_s, 可以得到如果 φ 是一个左积分, 那么 $\varphi \circ S$ 是一个右积分. 类似地, 如果 ψ 是一个右积分, 那么 $\psi \circ S$ 是一个右积分. 同样 $\varphi \circ S^2$ 是一个左积分如果 φ 是一个左积分. 但是, 因为积分不再是唯一的, 所以 $\varphi \circ S^2$ 不一定是 φ 的倍数.

命题 5.3.3 (i) 设 φ 是 A 上的线性泛函, 对任意 $a,b \in A$, 定义

$$c = (\mathbf{id} \otimes \varphi)(\Delta(a)(1 \otimes b)) \quad \text{和} \quad d = (\mathbf{id} \otimes \varphi)((1 \otimes a)\Delta(b)).$$

这些元素属于 A. 那么 φ 是左不变的当且仅当 $d = S(c)$ 对任意的 $a,b \in A$.

(ii) 类似地, 设 ψ 是 A 上的线性泛函, 对任意 $a,b \in A$, 定义

$$c = (\psi \otimes \mathbf{id})((a \otimes 1)\Delta(b)) \quad \text{和} \quad d = (\psi \otimes \mathbf{id})(\Delta(a)(b \otimes 1)).$$

那么 ψ 右不变的当且仅当 $d = S(c)$ 对任意的 $a,b \in A$.

证明 (i) (a) 首先假设 φ 是左不变的. 对任意的 $a,b \in A$, 有

$$(1 \otimes a)\Delta(b) = \sum (S(a_{(1)}) \otimes 1)\Delta(a_{(2)}b),$$

将 φ 作用在第二个因子可得

$$d = (\mathbf{id} \otimes \varphi)((1 \otimes a)\Delta(b)) = \sum S(a_{(1)})((\mathbf{id} \otimes \varphi)\Delta(a_{(2)}b)),$$

作用 Δ, 因为 $\Delta(px) = \Delta(p)(x \otimes 1)$, $p \in A, x \in A_t$, 从而可得

$$\Delta(d) = \sum \Delta(S(a_{(1)}))(((\mathbf{id} \otimes \varphi)\Delta(a_{(2)}b)) \otimes 1),$$

接下来在第二个因子从右端乘以元素 q, 再结合 S 在 Δ 上的作用可得

$$\Delta(d)(1 \otimes q) = \sum (S(a_{(2)}) \otimes S(a_{(1)})q)((\mathbf{id} \otimes \varphi)\Delta(a_{(3)}b) \otimes 1)$$
$$= \sum S(a_{(2)})((\mathbf{id} \otimes \varphi)\Delta(a_{(3)}b)) \otimes S(a_{(1)})q.$$

利用第一个公式并且将 a 替换为 $a_{(2)}$ 可得

$$\Delta(d)(1 \otimes q) = \sum (\mathbf{id} \otimes \varphi)((1 \otimes a_{(2)})\Delta(b)) \otimes S(a_{(1)})q.$$

现在将 ε 作用在上述等式的左边我们可以得到 $dq = S(c)q$, 这里 $c = \sum a_{(1)}\varphi(a_{(2)}b)$. 消去 q 从而就证明了命题的一部分.

(b) 反过来, 假设线性泛函 φ 满足要求. 对任意的 $p, q \in A$ 有

$$(\mathbf{id} \otimes \varphi)\Delta(pq) = \sum p_{(1)}(\mathbf{id} \otimes \varphi)((1 \otimes p_{(2)})\Delta(q))$$
$$= \sum p_{(1)} S((\mathbf{id} \otimes \varphi)(\Delta(p_{(2)})(1 \otimes q)))$$
$$= \sum p_{(1)} S(p_{(2)}) \varphi(p_{(3)} q).$$

如果我们从左侧乘以一个 A 中的元素, 那么上述式子中的元素都可以被覆盖. 因为 $A^2 = A$, 这就证明了 $(\mathbf{id} \otimes \varphi)\Delta(a) \in \varepsilon_t(A)$ 对任意的 $a \in A$. 又因为 $\varepsilon_t(A)$ 是 A_t 的子集, 从而 φ 是左不变的.

(ii) 对于右不变泛函的证明是类似的.

接下来我们将会证明一个命题, 对于定义 A 的对偶空间 $\hat{A}$ 有重要意义.

命题 5.3.4 设 φ 是 A 上的左积分, ψ 是 A 上的右积分. 对任意的 $p, q \in A$.
(i) 对任意的 $x \in A$ 如果

$$a = (\mathbf{id} \otimes \varphi)((\mathbf{id} \otimes S)(\Delta(p))(1 \otimes q)) \quad \text{和} \quad b = (\psi \otimes \mathbf{id})((S^{-1} \otimes \mathbf{id})(\Delta(q))(p \otimes 1)),$$

则 $\psi(xa) = \varphi(xb)$.

(ii) 对任意的 $x \in A$ 如果

$$a = (\mathbf{id} \otimes \varphi)((1 \otimes q)(\mathbf{id} \otimes S^{-1})\Delta(p)) \quad \text{和} \quad b = (\psi \otimes \mathbf{id})((p \otimes 1)(S \otimes \mathbf{id})\Delta(q)),$$

则 $\psi(ax) = \varphi(bx)$.

(iii) 对任意的 $x \in A$ 如果

$$a = (\mathbf{id} \otimes \varphi)((1 \otimes q)(\mathbf{id} \otimes S)\Delta(p)) \quad \text{和} \quad b = (\psi \otimes \mathbf{id})((S \otimes \mathbf{id})(\Delta(q))(p \otimes 1)),$$

则 $\psi(xa) = \varphi(bx)$.

(iv) 对任意的 $x \in A$ 如果

$$a = (\mathbf{id} \otimes \varphi)((\mathbf{id} \otimes S^{-1})(\Delta(p))(1 \otimes q)) \quad \text{和} \quad b = (\psi \otimes \mathbf{id})((p \otimes 1)(S^{-1} \otimes \mathbf{id})\Delta(q)),$$

则 $\psi(ax) = \varphi(xb)$.

证明 首先证明 (i), 令 $p, q, x \in A$. 考虑 $(\psi \otimes \varphi)(\Delta(x)(p \otimes q))$. 利用命题 5.3.3 关于 φ 的公式和将 $a, b \in A$ 替换为 p, q 可得

$$(\psi \otimes \varphi)(\Delta(x)(p \otimes q)) = \psi(\cdot p)((1 \otimes \varphi)(\Delta(x)(1 \otimes q)))$$
$$= \psi(\cdot p)(S^{-1}((1 \otimes \varphi)((1 \otimes x)\Delta(q))))$$
$$= (\psi \otimes \varphi)((1 \otimes x)(S^{-1} \otimes 1)\Delta(q)(p \otimes 1))$$
$$= \varphi(xb),$$

这里 $b = (\psi \otimes \mathbf{id})((S^{-1} \otimes \mathbf{id})\Delta(q)(p \otimes 1))$. 另一方面, 利用命题 5.3.3 关于 ψ 的公式, 可得

$$(\psi \otimes \varphi)(\Delta(x)(p \otimes q)) = \varphi(\cdot q)((\psi \otimes \mathbf{id})(\Delta(x)(p \otimes 1)))$$
$$= \varphi(\cdot q)(S((\psi \otimes \mathbf{id})(x \otimes 1)\Delta(p)))$$
$$= (\psi \otimes \varphi)((x \otimes 1)(\mathbf{id} \otimes S)\Delta(p)(1 \otimes q))$$
$$= \psi(xa),$$

这里 $a = (\mathbf{id} \otimes \varphi)((\mathbf{id} \otimes S)\Delta(p)(1 \otimes q))$. 这就证明了 (i).

接下来证明 (ii). 只要考虑 $(\psi \otimes \varphi)((p \otimes q)\Delta(x))$, 进行上述过程.

接下来证明 (iii), 只要考虑 $(\psi \otimes \varphi)((1 \otimes q)\Delta(x)(p \otimes 1))$. 对 (iv) 的证明只需考虑 $(\psi \otimes \varphi)((p \otimes 1)\Delta(x)(1 \otimes q))$.

定义 5.3.5 称一个弱乘子 Hopf 代数拥有一个忠实的集合的积分 (a faithful set of integrals) 如果满足下面两个条件. 对任意的 $x \in A$, 可以得到 $x = 0$, 如果 $\varphi(xa) = 0$ 对任意的左积分 φ 和 $a \in A$. 类似地, 如果 $\varphi(ax) = 0$ 对任意的左积分 φ 和 $a \in A$, 可以得到 $x = 0$.

定义 5.3.6 假设 (A, Δ) 是一个正则的弱乘子 Hopf 代数并带有一个忠实的集合的积分. 则可定义空间 $\widehat{A} = \{\varphi(\cdot a) | \varphi 是左积分, a \in A\}$.

在接下来这个命题中我们将说明 $\widehat{A}$ 代表元素的选取不影响定义.

命题 5.3.7 对任意左积分 φ 和右积分 ψ 以及 $a \in A$. 我们有四种形式的线性泛函 $\varphi(\cdot a), \varphi(a\cdot), \psi(\cdot a)$ 和 $\psi(a\cdot)$, 它们都属于 $\widehat{A}$. 并且 $\widehat{A}$ 是这四种形式之一的线性泛函的线性扩张, φ 取遍所有的左积分, ψ 取遍所有的右积分, a 取遍所有的 A 中的元素.

证明 将对极作用在所有的左积分上, 可得所有的右积分构成的集合是忠实的.

所有的 $\widehat{A}$ 中的元素都可以看成是命题 5.3.4 中八种元素形式之一的线性扩张. 接下来我们将证明对于下列形式是对的.

$$(\imath \otimes \varphi)((\imath \otimes S)(\Delta(p))(1 \otimes q)),$$

$p, q \in A$. 反证法. 假设这不正确, 那么就存在 A 上的一个非零的线性泛函 ω 作用在所有这种类型的元素均为 0. ω, 我们有 $\varphi(S(r)q) = 0$, $r = (\omega \otimes \imath)\Delta(p)$ 对任意的 $p, q \in A$ 以及对任意的左积分 φ. 将 q 替换为 qq' 并且利用积分的忠实性可得 $S(r)q = 0$ 对任意 $p, q \in A$. 又 S 是双射, 可得

$$(\omega \otimes \imath)((1 \otimes q)\Delta(p)) = 0$$

5.3 Pontryagin 对偶

对任意的 $p,q \in A$. 由于余乘是完全的, 可得 $\omega = 0$. 矛盾. 对于其他情形也可类似证明.

所以, 利用命题 5.3.4 的结果, 不仅得到了这四种类型的元素都属于 $\widehat{A}$, 并且还得出 $\widehat{A}$ 是四种形式之一的元素的线性扩张.

定义 5.3.8 假设 (A, Δ) 是一个正则的弱乘子 Hopf 代数并带有一个忠实的集合的积分, 那么我们称 A 为**代数量子群胚**(algebraic quantum groupoid) 或者**弱代数量子群**(weak algebraic quantum group).

接下来将假设 A 是一个代数量子群胚, 考虑 A 的对偶空间 $\widehat{A}$. 首先我们考虑 $\widehat{A}$ 上的乘法.

命题 5.3.9 对任意 $\omega, \omega' \in \widehat{A}$, 定义乘法 $\omega\omega' \in \widehat{A}$ 为
$$(\omega\omega')(x) = (\omega \otimes \omega')\Delta(x),$$
$x \in A$. 如果 $\omega' = \varphi(a \cdot)$, 这里 φ 是一个左积分, $a \in A$, 那么 $\omega\omega' = \varphi(b \cdot)$ 其中
$$b = ((\omega \circ S) \otimes \mathbf{id})\Delta(a).$$

证明 首先我们说明定义是良好的. 令 $\omega' = \sum_i \varphi_i(\cdot a_i)$, φ_i 是左积分, $a_i \in A$. 那么
$$(\omega \otimes \omega')\Delta(x) = \sum_i (\omega \otimes \varphi_i)(\Delta(x)(1 \otimes a_i))$$
对任意的 x 是定义良好的. 因为对任意的 a_i 都有 $\Delta(x)(1 \otimes a_i)$ 属于 $A \otimes A$. 接下来我们证明乘积 $\omega\omega'$ 不依赖于代表元素的选择. 假设 ω 等于 $\varphi(\cdot a)$.

接下来我们要证明 $\omega\omega'$ 仍然属于 $\widehat{A}$. 令 $\omega' = \varphi(a \cdot)$ 利用命题 5.3.3 中的公式可得
$$(\omega \otimes \omega')\Delta(x) = (\omega \otimes \varphi)((1 \otimes a)\Delta(x)) = ((\omega \circ S) \otimes \varphi)(\Delta(a)(1 \otimes x)).$$
从而 $\omega\omega' = \varphi(b \cdot)$, $b = ((\omega \circ S) \otimes 1)\Delta(a)$. 因为 $\omega \in \widehat{A}$ 可得 b 属于 A.

命题 5.3.10 当 $\widehat{A}$ 带有上述乘法时, $\widehat{A}$ 是一个结合的非退化的幂等的代数.
证明见文献 [202].

接下来将构造 $\widehat{A}$ 中的余乘, 为了构造余乘, 首先研究对偶对 $\langle A, \widehat{A} \rangle$. $\omega(a)$ 将会被记为 $\langle a, \omega \rangle$, $a \in A, \otimes \in \widehat{A}$. 将会用 B 替代 $\widehat{A}$. A 中的元素将会被记为 $a, a', \cdots$. B 中的元素将会被记为 $b, b', \cdots$.

命题 5.3.11 给定对偶对 $\langle A, B \rangle$ 存在如下四种作用:
$$\langle a'a, b \rangle = \langle a', a \triangleright b \rangle, \quad \langle a, bb' \rangle = \langle a \triangleleft b, b' \rangle,$$
$$\langle aa', b \rangle = \langle a', b \triangleleft a \rangle, \quad \langle a, b'b \rangle = \langle b \triangleright a, b' \rangle,$$

$a, a' \in A, b, b' \in B$. 在两种类型中,左右作用是互相交换的. 所有的四种作用都是单位的、非退化的.

接下来将会用上述命题的结果, 将对偶对从 $A \times B$ 扩张到 $A \times M(B)$ 和 $M(A \times B)$.

命题 5.3.12 对偶对可以按照如下方式从 $A \times B$ 扩张到 $A \times M(B)$ 和 $M(A \times B)$

$$\langle a, bm \rangle = \langle a \triangleleft b, m \rangle \quad 和 \quad \langle a, mb \rangle = \langle b \triangleright a, m \rangle,$$

$a \in A, b \in B, m \in M(A)$.

证明 令 $m \in M(B)$, $a \in A$, 因为 B 在 A 上的右作用 $\triangleleft$ 是单位的, $a = \sum_i a_i \triangleleft b_i$. 定义

$$\langle a, m \rangle = \sum_i \langle a_i, b_i m \rangle.$$

假设 $\sum_i a_i \triangleleft b_i = 0$. 对任意的 $b \in B$ 有

$$\sum_i \langle a_i \triangleleft (b_i m), b \rangle = \sum_i \langle a_i, b_i(mb) \rangle = \sum_i \langle a_i \triangleleft b_i, mb \rangle = 0.$$

由此可导出 $\sum_i a_i \triangleleft (b_i m) = 0$. 作用余单位 ε 可得

$$\sum_i \langle a_i, b_i m \rangle = 0.$$

这就表明定义是良好的. 对于第二个式子可以类似证明. 唯一性的证明是平凡的.

命题 5.3.13 存在一个 $\widehat{A}$ 上的正则的余乘 $\widehat{\Delta}$ 使得对应的典范映射 $\widehat{T_1}$ 和 $\widehat{T_2}$ 满足

$$\langle x \otimes y, \widehat{T_1}(\omega \otimes \omega') \rangle = \langle T_2(x \otimes y), \omega \otimes \omega' \rangle, \tag{5.3.1}$$

$$\langle x \otimes y, \widehat{T_2}(\omega \otimes \omega') \rangle = \langle T_1(x \otimes y), \omega \otimes \omega' \rangle \tag{5.3.2}$$

对任意的 $x, y \in A, \omega, \omega' \in \widehat{A}$.

证明见文献 [202]. 从证明过程知 Δ 是一个正则的余乘, 并且还是代数同态. 接下来我们将得到一个关于 Δ 的重要的公式.

命题 5.3.14 对任意的 $a, a' \in A, b \in B$, $\langle aa', b \rangle = \langle a \otimes a', \Delta(b) \rangle$.

证明 任取 $a, a' \in A, b \in B$, 有

$$\langle a \otimes (b' \triangleright a'), \Delta(b) \rangle = \langle a \otimes a', \Delta(b)(1 \otimes b') \rangle$$
$$= \langle (a \otimes 1)\Delta(a'), b \otimes b' \rangle$$
$$= \langle a(b' \triangleright a'), b \rangle.$$

5.3 Pontryagin 对偶

因为 B 在 A 上的作用是单位的, 结合上面等式可以得出 $\langle aa', b\rangle = \langle a\otimes a', \Delta(b)\rangle$.

命题 5.3.15 余乘 $\widehat{\Delta}$ 是完全的.

证明 由命题 5.3.13 的证明有

$$\widehat{\Delta}(\omega)(1\otimes\omega') = \sum \omega(\cdot S(a_{(1)})) \otimes \varphi(a_{(2)}\cdot),$$

$\omega, \omega' \in \widehat{A}$, $\omega' = \varphi(a\cdot)$, $a \in A$, φ 是 A 上的一个左积分. 设 $\gamma \in \widehat{A}$ 在 $c \in A$ 处取值, 可得

$$(1\otimes\gamma(\cdot\omega'))\widehat{\Delta}(\omega) = \sum \omega(\cdot a_{(1)}) \otimes \varphi(a_{(2)}c).$$

为了要证明 $\widehat{\Delta}$ 的左腿是 $\widehat{A}$, 只需证明 A 中的元素是下列形式的元素的线性组合,

$$\sum S(a_{(1)})\varphi(a_{(2)}c).$$

反证法, 假设这是错误的. 那么就存在一个 A 上的线性泛函 ρ 满足

$$\sum \rho(S(a_{(1)}))\varphi(a_{(2)}c) = 0.$$

因为 A 上的积分是忠实的, 可得

$$\sum \rho(S(a_{(1)}))a_{(2)} = 0$$

对任意的 a. 又因为原余乘是完全的, 由此可得 $\rho = 0$. 这就证明了 $\widehat{\Delta}$ 的左腿是 $\widehat{A}$. 类似地, 可以证明 $\widehat{\Delta}$ 的右腿是 $\widehat{A}$.

命题 5.3.16 在 $\widehat{A}$ 上存在一个余单位 $\widehat{\varepsilon}$ 定义为 $\widehat{\varepsilon}(\omega) = \omega(1)$.

证明 在这里应用命题 5.3.12 来定义 $\omega(1), \omega \in \widehat{A}$.

根据定义, 当 $\omega = \varphi(\cdot a)$ 时, 有 $\widehat{\varepsilon}(\omega(\cdot a)) = \omega(a)$ 和 $\widehat{\varepsilon}(\omega(a\cdot)) = \omega(a)$ 对任意的 $a \in A$ 和 $\omega \in \widehat{A}$. 考虑 $\omega, \omega' \in \widehat{A}$. 根据定义有

$$(\widehat{\Delta}(\omega)(1\otimes\omega'))(a\otimes a') = (\omega\otimes\omega')((a\otimes 1)\Delta(a')).$$

因此可以推出

$$(\widehat{\varepsilon}\otimes\mathbf{id})(\widehat{\Delta}(\omega)(1\otimes\omega')) = (\omega\otimes\omega')\circ\Delta = \omega\omega'.$$

类似地, 有

$$(\mathbf{id}\otimes\widehat{\varepsilon})((\omega\otimes 1)\widehat{\Delta}(\omega)) = \omega\omega'.$$

这就证明了 $\widehat{\varepsilon}$ 是余单位. 又余乘是完全的, 所以这个余单位是唯一的.

接下来考虑 $\widehat{A}$ 上的反对极 $\widehat{S}$.

命题 5.3.17 存在一个映射 $\widehat{S}: \widehat{A} \longrightarrow \widehat{A}$ 满足

$$\langle a, \widehat{S}(b)\rangle = \langle S(a), b\rangle$$

对任意的 $a \in A, b \in \widehat{A}$. $\widehat{S}$ 是双射.

证明 上述公式可以用来定义 $\widehat{S}(b) \in A'$ 对任意的 $b \in A'$. 我们只需证明 $\widehat{S}(b) \in \widehat{A}$ 对任意的 $b \in \widehat{A}$.

令 $b = \varphi(\cdot c), c \in A, \varphi$ 是 A 上的一个左积分. 那么

$$\langle a, \widehat{S}(b) \rangle = \varphi(S(a)c) = (\varphi \circ S)(S^{-1}(c)a),$$

这里 $a \in A$. 从而 $\widehat{S}(b) = (\varphi \circ S)(S^{-1}(c) \cdot)$. 我们知道 $\varphi \circ S$ 是一个右积分. 由命题 5.3.7 可得 $\widehat{S}(b) \in \widehat{A}$.

从上述证明可以看出 $\widehat{S}$ 是一个从 $\widehat{A}$ 到它自己的双射.

命题 5.3.18 存在映射 $\widehat{R_1}$ 和 $\widehat{R_2}$ 从 $\widehat{A} \otimes \widehat{A}$ 到它自己, 满足

$$\langle a \otimes a', \widehat{R_1}(b \otimes b') \rangle = \langle R_2(a \otimes a'), b \otimes b' \rangle, \tag{5.3.3}$$

$$\langle a \otimes a', \widehat{R_2}(b \otimes b') \rangle = \langle R_1(a \otimes a'), b \otimes b' \rangle \tag{5.3.4}$$

对任意的 $a, a' \in A, b, b' \in B$. 并且有

$$\widehat{R_1}(b \otimes b') = \sum b_{(1)} \otimes \widehat{S}(b_{(2)})b', \tag{5.3.5}$$

$$\widehat{R_2}(b \otimes b') = \sum b\widehat{S}(b'_{(1)}) \otimes b'_{(2)}, \tag{5.3.6}$$

$\widehat{R_1}$ 和 $\widehat{R_2}$ 分别是 $\widehat{T_1}$ 和 $\widehat{T_2}$ 的广义逆.

证明 首先可以利用公式 (5.3.3) 和 (5.3.4) 来定义 $\widehat{R_1}(b \otimes b')$ 和 $\widehat{R_2}(b \otimes b')$ 属于 $(A \otimes A)'$, 对任意的 $b, b' \in A'$. 类似于命题 5.3.13 的证明这些元素属于 $\widehat{A} \otimes \widehat{A}$ 如果 $b, b' \in \widehat{A}$. 事实上, 我们也可以利用 $\widehat{S}$ 是一个反自同构再结合 $\widehat{A} \otimes \widehat{A}$ 上的典范映射的值域为 $\widehat{A} \otimes \widehat{A}$. 这就证明了命题的第一部分.

接下来证明等式 (5.3.5). 等式 (5.3.6) 的证明类似. 任取 $a, a' \in A, b, b' \in \widehat{A}$. 有

$$\begin{aligned}\langle a \otimes a', \widehat{R_1}(b \otimes b') \rangle &= \langle a \otimes \Delta(a'), (1 \otimes \widehat{S} \otimes \mathbf{id})(\widehat{\Delta}(b) \otimes b') \rangle \\ &= \langle a \otimes (S \otimes \mathbf{id})\Delta(a'), \widehat{\Delta}(b) \otimes b' \rangle \\ &= \langle (a \otimes 1)(S \otimes \mathbf{id})\Delta(a'), b \otimes b' \rangle \\ &= \langle R_2(a \otimes a'), b \otimes b' \rangle.\end{aligned}$$

这就证明了等式 (5.3.5).

最后, 不难证明 $\widehat{R_1}$ 是 $\widehat{T_1}$ 的广义逆, 因为它们是 R_2 和 T_2 的伴随, 而 R_2 和 T_2 又是互为广义逆. 类似地, $\widehat{R_2}$ 是 $\widehat{T_2}$ 的广义逆.

为了符号简洁, 在接下来的内容中我们将会用 Δ 来表示对偶余乘, 不再用 $\widehat{\Delta}$. 对于典范映射和对应的广义逆也采取同样的记法. 接下来我们定义典范幂等元 $\widehat{E}$.

5.3 Pontryagin 对偶

命题 5.3.19 存在一个乘子 $\widehat{E}$ 属于 $M(B \otimes B)$ 定义为

$$\langle a \otimes a', \widehat{E} \rangle = \varepsilon(aa')$$

对任意的 $a, a' \in A$. 它满足

$$\widehat{E}(b \otimes b') = T_1 R_1(b \otimes b') \quad \text{和} \quad (b \otimes b')\widehat{E} = T_2 R_2(b \otimes b')$$

对任意的 $b, b' \in B$. 特别地, 它是一个幂等元并且决定了典范映射的值域.

证明 首先利用上述式子定义 $\widehat{E} \in (A \otimes A)'$. 任取 $b, b' \in \widehat{A}$. 利用 $(A \otimes A)'$ 中的乘法, 可得

$$\langle a \otimes a', \widehat{E}(b \otimes b') \rangle = \sum \varepsilon(a_{(1)} a'_{(1)}) \langle a_{(2)}, b \rangle \langle a'_{(2)}, b' \rangle. \tag{5.3.7}$$

现在可得

$$\sum \Delta(a)(a'_{(1)} \otimes 1) \otimes a'_{(2)} = \sum \Delta(aa'_{(1)})(1 \otimes S(a'_{(2)})) \otimes a'_{(3)},$$

将 ε 作用在第一个因子上, 有

$$\sum \varepsilon(a_{(1)} a'_{(1)}) a_{(2)} \otimes a'_{(2)} = \sum aa'_{(1)} S(a'_{(2)}) \otimes a'_{(3)} = R_2 T_2(a \otimes a').$$

如果将这应用在等式 (5.3.7), 可得

$$\langle a \otimes a', \widehat{E}(b \otimes b') \rangle = \langle R_2 T_2(a \otimes a'), b \otimes b' \rangle$$

对任意的 a, a', b, b'. 这就证明了 $\widehat{E}(b \otimes b') = T_1 R_1(b \otimes b')$. 类似地, 可以证明 $(b \otimes b')\widehat{E} = T_2 R_2(b \otimes b')$.

命题 5.3.20 对于上面构造的对偶余乘 Δ 以及典范幂等元 $\widehat{E}$, 有

$$(\Delta \otimes \mathbf{id})\widehat{E} = (\widehat{E} \otimes 1)(1 \otimes \widehat{E}) \quad \text{和} \quad (\Delta \otimes \mathbf{id})\widehat{E} = (1 \otimes \widehat{E})(\widehat{E} \otimes 1).$$

证明 为了证明第一个等式, 从左边乘以 $\Delta(b) \otimes b'$, $b, b' \in B$. 那么它等价于证明

$$(\Delta \otimes 1)T_2 R_2 = (1 \otimes T_2 R_2)(\Delta \otimes 1).$$

这个等式是下列式子的伴随

$$R_1 T_1(aa' \otimes a'') = (a \otimes 1) R_1 T_1(a' \otimes a'')$$

对任意的 $a, a', a'' \in A$. 又因为 $R_1 T_1(a \otimes a') = (a \otimes 1) F_1(1 \otimes a')$ 对任意的 $a, a' \in A$, $F_1 = (\mathbf{id} \otimes S)E$ (见 [202] 中命题 4.5 和命题 4.7), 从而等式成立.

为了证明第二个等式从右边乘以 $\Delta(b) \otimes b'$. 那么就等价证明下列式子成立.

$$(\Delta \otimes \mathbf{id})T_1 R_1 = (1 \otimes T_1 R_1)(\Delta \otimes \mathbf{id}).$$

而它是下列等式的伴随

$$R_2 T_2 (aa' \otimes a'') = (a \otimes 1) R_2 T_2 (a' \otimes a'')$$

对任意的 $a, a', a'' \in A$. 再次利用 [201] 中的命题 4.5 和命题 4.7 可知等式成立.

接下来我们可以得到本节的主要定理.

定理 5.3.21 对偶对 $(\widehat{A}, \widehat{\Delta})$ 仍然是一个弱乘子 Hopf 代数.

证明 根据上述构造可得定理 5.1.4 所有的条件均满足,从而可知 $\widehat{A}$ 是一个弱乘子 Hopf 代数. 又因为 $\widehat{S}$ 是它的对极并且 $\widehat{S}$ 是双射,所以根据定理 5.1.6 可知 $\widehat{A}$ 是正则的.

接下来我们将证明对偶弱乘子 Hopf 代数上也存在正则的积分. 首先考虑存在性.

命题 5.3.22 对任意的 $a \in A$, 存在一个 B 上的右不变积分 ψ_a 满足

$$\psi_a(\omega) = \varphi(a\varepsilon_s(c)), \qquad 当 \quad \omega = \varphi(\cdot c), \tag{5.3.8}$$

这里 $c \in A$, φ 是 A 上的左积分.

证明 令 $a \in A$.

(i) 首先我们证明可以定义 ψ_a 使得 (5.3.8) 成立. 假设存在有限个左积分 φ_i 使得

$$\sum_i \varphi_i(\cdot c_i) = 0$$

对任意的 $c_i \in A$. 这就可导出

$$\sum_i (1 \otimes \varphi_i)(\Delta(x)(1 \otimes c_i)) = 0$$

对任意的 $x \in A$. 利用命题 5.3.4 中的公式可得

$$\sum_i (1 \otimes \varphi_i)((1 \otimes x)\Delta(c_i)) = 0$$

对任意的 x. 使用 Sweedler 记号,可得

$$\sum_i aS(c_{i(1)}) \otimes \varphi_i(\cdot c_{i(2)}) = 0.$$

5.3 Pontryagin 对偶

对这个等式作用赋值映射 $x \otimes \omega \mapsto \omega(x)$，可得

$$\sum_i \varphi_i(aS(c_{i(1)})c_{i(2)}) = \sum_i \varphi_i(a\varepsilon_s(c_i)) = 0.$$

这就表明我们可以定义一个 B 上的线性映射 ψ_a 满足 (5.3.8)。

(ii) 接下来我们将证明 ψ_a 在 B 上是右不变的。设 $c, c' \in A$，φ 是 A 上的左积分。令 $\omega = \varphi(\cdot c)$。那么有

$$\psi_a(\omega(\cdot c')) = \varphi(a\varepsilon_s(c'c)) = \varphi(a\varepsilon_s(\varepsilon_s(c')c)),$$

这里 $\varepsilon_s(c'c) = \varepsilon_s(\varepsilon_s(c')c)$。由此可得

$$\psi_a(\omega(\cdot c')) = \psi_a(\varphi(\cdot \varepsilon_s(c')c)) = \psi_a(\omega(\cdot \varepsilon_s(c'))).$$

利用 Sweedler 记号，对偶记号及将 ω 记为 b，可得

$$\sum \psi_a(b_{(1)})\langle c', b_{(2)}\rangle = \sum \psi_a(b_{(1)})\langle \varepsilon_s(c'), b_{(2)}\rangle$$
$$= \sum \psi_a(b_{(1)})\langle c', \varepsilon_s(b_{(2)})\rangle.$$

在上述等式证明中我们用到了靶映射是自对偶的。这就表明

$$\sum \psi_a(b_{(1)})b_{(2)} = \sum \psi_a(b_{(1)})\varepsilon_s(b_{(2)}),$$

最终我们证明了 ψ_a 的右不变性。

现在我们将证明上述命题中构造的积分都是正则的。

命题 5.3.23 上述积分所构成的集合是正则的。

证明 (i) 设 $\omega \in B$，假设 $\psi_a(\omega\omega') = 0$ 对任意的 $a \in A, \omega' \in B$。我们将要证明 $\omega = 0$。如果 $\omega' = \varphi(\cdot c')$，那么 $\omega\omega' = \varphi(\cdot c)$，$c = (\omega \circ S^{-1} \otimes 1)\Delta(c')$。由此可得 $\psi_a(\omega\omega') = \varphi(a\varepsilon_s(c))$。又

$$\varepsilon_s(c) = \sum \omega(S^{-1}(c'_{(1)})\varepsilon(c'_{(2)}))$$

使得

$$\psi_a(\omega\omega') = \sum \varphi(a\varepsilon_s(c'_{(2)}))\omega(S^{-1}(c'_{(1)})).$$

根据假设，可得

$$\sum \omega(S^{-1}(c'_{(1)})\varepsilon_s(c'_{(2)})) = 0$$

对任意的 c' 都成立，对上式从右边乘以 $S(c'_{(3)})$ 可得

$$\sum \omega(S^{-1}(c'_{(1)})S(c'_{(2)})) = 0.$$

接下来两边都作用余单位可得 $\omega = 0$.

(ii) 令 $\omega' \in B$, 假设 $\psi_a(\omega\omega') = 0$ 对任意的 $a \in A, \omega' \in B$. 我们需要证明 $\omega' = 0$. 类似 (i) 的证明可以得到

$$\sum \omega(S^{-1}(c'_{(1)})S(c'_{(2)})) = 0$$

对任意的 ω. 因此 $\omega' = 0$.

现在我们得到 Pontryagin 对偶定理.

定理 5.3.24 一个带有足够多积分的弱乘子 Hopf 代数 (A, Δ) 的对偶同样有一组积分集, 并且可知 $(\widehat{A}, \widehat{\Delta})$ 的对偶与 (A, Δ) 之间存在一个典范同构.

5.4 乘子 Hopf 代数胚

首先我们给出仅在本节中使用的符号的说明. 集合 X 上的恒等映射我们将会记为 id_X 或者 id. 对于 B 上的右模 M, 我们将记为 M_B. 用 B_B 或者 $_BB$ 表示 B 是一个右 B 模或者左 B 模, 其中模作用分别是右乘或者左乘. 类似于前面章节, $L(B)$ 和 $R(B)$ 分别表示 B 的左乘子和右乘子集合. 记 X 的线性扩张为 $\mathrm{span}\, X$.

设 A 是一个非退化的幂等代数, 不一定有单位, 用 m 表示乘法, 用 m^{op} 表示反代数 A^{op} 中的乘法. 设 C 是一个不一定有单位的代数, 并且有同态 $s: C \to M(A)$ 和反同态 $t: C \to M(A)$ 使得 $s(C)$ 与 $t(C)$ 交换. 记 C 中的元素为 $x, x', y, y', \cdots$. 用 $_CA$ 和 A^C 分别表示在作用 $x \cdot a = s(x)a, a \cdot x = t(x)a$ 下构成的左 C 模和右 C 模. 类似地, 有 A_C 和 ^CA.

记子空间
$$A^C \overline{\times}_C A \subset \mathrm{End}(A^C \otimes_C A)$$
是由满足下列条件的自同态 T 构成的: 对任意的 $a, b \in A$, 存在元素
$$T(a \otimes 1) \in A^C \otimes_C A \quad \text{和} \quad T(1 \otimes b) \in A^C \otimes_C A$$
满足
$$T(a \otimes b) = (T(a \otimes 1))(1 \otimes b) = (T(1 \otimes b))(a \otimes 1).$$

如果 A 有单位 1_A, 那么 $A^C \overline{\times}_C A$ 到 $A^C \otimes CA$ 之间有一个嵌入, 嵌入映射为 $T \mapsto T(1_A \otimes 1_A)$. $A^C \overline{\times}_C A$ 表示左 Takeuchi 乘法代数:
$$\{w \in A^C \otimes_C A : w(t(x) \otimes 1) = w(1 \otimes s(x)), \forall x \in C\}.$$

设 B 是一个代数并带有同态 $t: B \to M(A)$ 和反同态 $t: B \to M(A)$ 使得 $s(B)$ 与 $t(B)$ 交换. 在作用 $a \cdot y = as(y)$ 和 $y \cdot a = at(y)$ 下有模 A_B 和 ^BA, 对任意的

5.4 乘子 Hopf 代数胚

$a \in A, b \in B$. 类似地, 也可以定义模 ${}_B A$ 和 A^B. 记

$$A_B \overline{\times}{}^B A \subset \text{End}(A_B \otimes {}^B A)^{\text{op}}.$$

这个子空间是由所有的满足下列条件的自同态 T 构成的:

$$(a \otimes b)T = (1 \otimes b)((a \otimes 1)T) = (a \otimes 1)((1 \otimes b)T).$$

该子空间是一个子代数. 如果 A 有单位, 那么映射 $T \to (1_A \otimes 1_A)T$ 给出了一个嵌入到带有右 Takeuchi 乘法的代数:

$$\{w \in A_B \otimes {}^B A : (s(y) \otimes 1)w = (1 \otimes t(y))w, \forall y \in B\}.$$

引理 5.4.1 设 $\Delta : A \to A^C \overline{\times}_C A$ 是线性映射. 定义映射 $\widetilde{T_\lambda}, \widetilde{T_\rho} : A \otimes A \to A^C \otimes CA$ 如下

$$\widetilde{T_\lambda}(a \otimes b) = \Delta(b)(a \otimes 1), \quad \widetilde{T_\rho}(a \otimes b) = \Delta(a)(1 \otimes b)$$

对任意的 $a, b \in A$. 并且这两个映射满足

$$\widetilde{T_\lambda}(t(x)a \otimes b) = \widetilde{T_\lambda}(a \otimes b)(1 \otimes s(x)), \quad \widetilde{T_\rho}(a \otimes s(y)b) = \widetilde{T_\rho}(a \otimes b)(t(y) \otimes 1),$$

$$(\mathbf{id} \otimes m)(\widetilde{T_\lambda} \otimes \mathbf{id}) = (m^{\text{op}} \otimes \mathbf{id})(\mathbf{id} \otimes \widetilde{T_\rho}),$$

$$(m^{\text{op}} \otimes \mathbf{id})(\mathbf{id} \otimes \widetilde{T_\lambda}) = \widetilde{T_\lambda}(m^{\text{op}} \otimes \mathbf{id}), \quad (\mathbf{id} \otimes m)(\widetilde{T_\rho} \otimes \mathbf{id}) = \widetilde{T_\rho}(\mathbf{id} \otimes m)$$

对任意的 $a, b \in A, x, y \in B$.

定义 5.4.2 一个**左乘子双代数胚**(left multiplier Hopf algebroid) 是一个五元组 $\mathcal{A} = (A, C, s, t, \Delta)$, 其中

(i) A 和 C 为代数且使得右 A-模 A_A 是非退化的和幂等的;

(ii) 同态 $s : C \to M(A)$ 和反同态 $t : C \to M(A)$ 的像可交换, 且使得 C-模 ${}_C A$ 和 C-模 A^C 是忠实和幂等的, 且 $A^C \otimes_C A$ 作为 $A \otimes 1$ 和 $1 \otimes A$ 上的右模是非退化的;

(iii) 同态 $\Delta : A \to A^C \overline{\times}_C A$ 称为左余乘, 其中 Δ 满足如下的 C-双线性条件 (5.4.1) 和余结合条件 (5.4.2).

$$\Delta(s(y)t(x)as(y')t(x')) = (s(y) \otimes t(x))\Delta(a)(s(y') \otimes t(x')); \tag{5.4.1}$$

$$(\Delta \otimes \iota)(\Delta(b)(1 \otimes c))(a \otimes 1 \otimes 1) = (\iota \otimes \Delta)(\Delta(b)(a \otimes 1))(1 \otimes 1 \otimes c), \tag{5.4.2}$$

其中 $a, b, c \in A$, $x, y, x', y' \in C$.

我们称映射 T_λ 和 T_ρ 为 $\mathcal{A}$ 的典范映射. 一个左乘子双代数胚称为**单位的**, 如果 A 和 C 以及映射 s, t, Δ 是**单位的**.

下面我们将给出左乘子双代数胚的左余单位的定义.

定义 5.4.3 一个左乘子双代数胚 (A, C, s, t, Δ) 的**左余单位**是一个映射 $\varepsilon : A \to C$ 满足:

$$\varepsilon(s(y)a) = y\varepsilon(a) \quad 和 \quad \varepsilon(t(x)a) = \varepsilon(a)x \quad 对所有的 a \in A, x, y \in C, \quad (5.4.3)$$

也就是说, $\varepsilon \in \mathrm{Hom}(_C A, _C C) \cap \mathrm{Hom}(A^C, C_C)$ 且满足:

$$(\varepsilon \otimes \iota)(T_\rho(a \otimes b)) = ab \quad 和 \quad (\iota \otimes \varepsilon)(T_\lambda(a \otimes b)) = ba \quad 对所有的 a, b \in A. \quad (5.4.4)$$

对于一个左乘子双代数胚, 我们也给出其完全的定义如下.

定义 5.4.4 我们称一个左乘子双代数胚 (A, C, s, t, Δ) 是**左完全的**, 当且仅当 A 是等价于由形式为 $(\iota \otimes \psi)(\widetilde{T}_\rho(a \otimes b))$ 的元素线性扩张生成的空间, 其中 $\psi \in \mathrm{Hom}(_C A, _C C), a, b \in A$; 称为**右完全的**, 当且仅当 A 是等价于由形式为 $(\phi \otimes \iota)(\widetilde{T}_\lambda(a \otimes b))$ 的元素线性扩张生成的空间, 其中 $\phi \in \mathrm{Hom}(A^C, C_C), a, b \in A$; 称为**完全的**当且仅当其既是左完全的又是右完全的.

注记 5.4.5 如果一个左乘子双代数胚 (A, C, s, t, Δ) 有左余单位 ε 且其典范映射 T_λ(或者 T_ρ) 是满射, 则其为左完全的 (或者右完全的). 要想得到这个结论, 可以取 ϕ(或者 ψ) 为 ε, 且 $AA = A$.

命题 5.4.6 设 $\mathcal{A} = (A, C, s, t, \Delta)$ 是一个左乘子双代数胚且含有左余单位 ε.
(i) 如果 $\mathcal{A}$ 是左完全的或者右完全的, 则左余单位是唯一的.
(ii) 如果典范映射 T_λ(或者 T_ρ) 是满射, 则对所有的 $a, b \in A$, 有

$$\varepsilon(ab) = \varepsilon(as(\varepsilon(b))) \quad (或者 \varepsilon(ab) = \varepsilon(at(\varepsilon(b)))). \quad (5.4.5)$$

证明 我们只证明结论 (1), 如果 $\mathcal{A}$ 是左完全的且 ε 是左余单位. 设 $a, b \in A, \psi \in \mathrm{Hom}(_C A, _C C)$, 且 $T_\rho(a \otimes b) = \sum_i c_i \otimes d_i$, 其中 $c_i, d_i \in A$. 则由 (5.4.4) 知, $\sum_i s(\varepsilon(c_i))d_i = ab$, 因此

$$\varepsilon\left(\sum_i t(\psi(d_i))c_i\right) = \sum_i \varepsilon(c_i)\psi(d_i) = \sum_i \psi(s(\varepsilon(c_i))d_i) = \psi(ab).$$

但由于 (A, C, s, t, Δ) 是左完全的, 即 $\sum_i t(\psi(d_i))c_i$ 扩张成 A. 如果 (A, C, s, t, Δ) 是右完全的, 则也可类似得到结论.

相对应地, 也可定义右乘子双代数胚.

定义 5.4.7 一个**右乘子双代数胚**(right multiplier Hopf algebroid) 是一个五元组 (A, B, s, t, Δ), 其中

5.4 乘子 Hopf 代数胚

(i) A 和 B 为代数且使得左 A-模 $_AA$ 是非退化的和幂等的;

(ii) 同态 $s: B \to M(A) \subseteq R(A)$ 和反同态 $t: B \to M(A) \subseteq R(A)$ 的像可交换, 且使得 B-模 A_B 和 B-模 ^BA 是忠实和幂等的, 且 $A_B \otimes^B A$ 作为 $A \otimes 1$ 和 $1 \otimes A$ 上的左模是非退化的;

(iii) 同态 $\Delta: A \to A_B \bar{\times}^B A$ 称为右余乘, 其中 Δ 满足:

$$\Delta(t(y)s(x)at(y')s(x')) = (t(y) \otimes s(x))\Delta(a)(t(y') \otimes s(x')), \tag{5.4.6}$$

$$(a \otimes 1 \otimes 1)((\Delta \otimes \iota)((1 \otimes c)\Delta(b))) = (1 \otimes 1 \otimes c)((\iota \otimes \Delta)((a \otimes 1)\Delta(b))), \tag{5.4.7}$$

其中 $a, b, c \in A$, $x, y \in B$.

一个右乘子双代数胚称为**单位的**, 如果 A 和 B 以及映射 s, t, Δ 是**单位的**.

和左乘子双代数胚相一致的, 也可以定义右乘子双代数胚的右余单位.

定义 5.4.8 一个右乘子双代数胚 $\mathcal{A} = (A, B, s, t, \Delta)$ 的**右余单位**(right counit) 是一个映射 $\varepsilon: A \to B$ 满足:

$$\varepsilon(at(y)) = y\varepsilon(a) \quad \text{和} \quad \varepsilon(as(x)) = \varepsilon(a)x \quad \text{对所有的} a \in A, \ x, y \in B, \tag{5.4.8}$$

也就是说, $\varepsilon \in \text{Hom}(A_B, B_B) \cap \text{Hom}(^BA, _BB)$ 且满足:

$$(\varepsilon \otimes \iota)(_\rho T(a \otimes b)) = ba \quad \text{和} \quad (\iota \otimes \varepsilon)(_\lambda T(a \otimes b)) = ab \quad \text{对所有的} a, b \in A. \tag{5.4.9}$$

和左乘子双代数胚相对应的, 对于右乘子双代数胚 $\mathcal{A} = (A, B, s, t, \Delta)$, 有:

(i) 类似于左乘子双代数胚可以定义右乘子双代数胚 $\mathcal{A}$ 的左完全性和右完全性, 且如果 $\mathcal{A}$ 是左完全的或者右完全的, 则 $\mathcal{A}$ 的右余单位是唯一的;

(ii) 如果映射 $_\rho T$(或者 $_\lambda T$) 是满射, 则

$$\varepsilon(ab) = \varepsilon(s(\varepsilon(a))b) \quad (\text{或者} \varepsilon(ab) = \varepsilon(t(\varepsilon(a))b)); \tag{5.4.10}$$

(iii) 不失一般性, 可以假设 ε 是满射;

(iv) 如果 $(_\rho T, _\lambda T)$ 是双射且 $A = As(^tI) = At(^sI)$, 其中 $^sI, ^tI \subseteq B$ 分别表示由所有的 $\phi \in \text{Hom}(A_B, B_B)$ 和 $\psi \in \text{Hom}(^BA, _BB)$ 的像线性张成的, 则 $\mathcal{A}$ 有唯一的右余单位.

接下来, 如果左乘子双代数胚和右乘子双代数胚的定义能够相对应地结合起来, 我们就可以开始定义双边的乘子双代数胚. 即, 一个乘子双代数胚由一个左乘子双代数胚 $\mathcal{A}_C = (A, C, s_C, t_C, \Delta_C)$ 和一个右乘子双代数胚 $\mathcal{A}_B = (A, B, s_B, t_B, \Delta_B)$ 满足下列设定:

首先, $\mathcal{A}_C$ 和 $\mathcal{A}_B$ 要有相同的底完全代数 A. 由前面的设定可知, A 作为代数是左右非退化的, 所以可以构建一个双边的乘子代数 $M(A)$ 来作为映射 s_B, t_B, s_C, t_C 的像空间.

第二个设定是为了第三个设定有意义, 有

$$s_B(B) = t_C(C), \qquad t_B(B) = s_C(C). \tag{5.4.11}$$

则有映射 $S_B := t_C^{-1} \circ s_B : B \to C$ 和 $S_C := t_B^{-1} \circ s_C : C \to B$ 均为反同构, 但是彼此不需要互逆. 为了简化记号, 用 B 表示 $s_B(B)$, 用 C 表示 $s_C(C)$, 也就是说, $s_B = \iota_B$, $s_C = \iota_C$, 则 $S_B = t_C^{-1}$, $S_C = t_B^{-1}$. 进一步地, B 中的元素表示成 $x, x', x'', \cdots$, C 中的元素表示成 $y, y', y'', \cdots$, 如果将 A 看成 B 的左模和右模 (模作用是乘法), 将 A 分别记成 $_BA$ 和 A_B, 如果将 A 看成 B 的右模和左模 (模作用分别是 $a \cdot x = t_B(x)a$ 和 $x \cdot a = at_B(x)$), 将 A 分别记成 A^B 和 ^BA. 类似地, A 作为 C 模可以分别记为 $_CA$, A_C, A^C 和 ^CA.

为了公式化第三个设定, 我们需要前面左乘子双代数胚和右乘子双代数胚的定义中 Δ_C 和 Δ_B 的 C-双线性性质和 B-双线性性质, 即式 (5.4.1) 和 (5.4.6), 此时可以写成

$$\Delta_C(xyax'y') = (y \otimes x)\Delta_C(a)(y' \otimes x'),$$
$$\Delta_B(xyax'y') = (y \otimes x)\Delta_B(a)(y' \otimes x') \tag{5.4.12}$$

对所有的 $a \in A, x, x' \in B, y, y' \in C$. 类似地, 对于余结合性, 可以得到如下混合余结合性:

$$((\Delta_C \otimes \iota)((1 \otimes c)\Delta_B(b)))(a \otimes 1 \otimes 1) = (1 \otimes 1 \otimes c)((\iota \otimes \Delta_B)(\Delta_C(b)(a \otimes 1))),$$
$$(a \otimes 1 \otimes 1)((\Delta_B \otimes \iota)(\Delta_C(b)(1 \otimes c))) = ((\iota \otimes \Delta_C)((a \otimes 1)\Delta_B(b)))(1 \otimes 1 \otimes c)$$
$$\tag{5.4.13}$$

对所有的 $a, b, c \in A$.

下面给出乘子双代数胚的定义.

定义 5.4.9 一个**乘子双代数胚**(multiplier bialgebroid)$\mathcal{A} = (A, B, C, t_B, t_C, \Delta_B, \Delta_C)$ 由下列组成:

(i) 一个非退化、幂等的代数 A,

(ii) 乘子代数 $M(A)$ 的子代数 B, C, 以及反同构 $t_B : B \to C$ 和 $t_C : C \to B$,

(iii) 映射 $\Delta_C : A \to A^C \bar{\times}_C A$ 和映射 $\Delta_B : A \to A_B \bar{\times}^B A$ 使得下面条件成立.

(iv) $\mathcal{A}_B = (A, B, \iota_B, t_B, \Delta_B)$ 是右乘子双代数胚,

(v) $\mathcal{A}_C = (A, C, \iota_C, t_C, \Delta_C)$ 是左乘子双代数胚,

(vi) 混合余结合性条件 (5.4.13) 成立.

我们称 $\mathcal{A}_C$ 的左余单位和 $\mathcal{A}_B$ 的右余单位分别为 $\mathcal{A}$ 的左余单位和右余单位. 同样地, $\mathcal{A}_C$ 的典范映射 T_λ 和 T_ρ 以及 $\mathcal{A}_B$ 的典范映射 $_\lambda T$ 和 $_\rho T$ 作为 $\mathcal{A}$ 的典范映射.

5.4 乘子 Hopf 代数胚

设 D 是一个代数, M 是 D 上的模, 记 M 的对偶为 $M^\vee := \mathrm{Hom}(M, D)$. 给一个乘子双代数胚 $(A, B, C, t_B, t_C, \Delta_B, \Delta_C)$, 定义如下子空间:

$$I_B := \mathrm{span}\{\omega(a) : \omega \in (A_B)^\vee, a \in A\}, \quad {}^BI := \mathrm{span}\{\omega(a) : \omega({}^BA)^\vee, a \in A\},$$

$$_CI := \mathrm{span}\{\omega(a) : \omega \in (_CA)^\vee, a \in A\}, \quad I^C := \mathrm{span}\{\omega(a) : \omega \in (A^C)^\vee, a \in A\}.$$

定义 5.4.10 称一个乘子双代数胚 $\mathcal{A} = (A, B, C, t_B, t_C, \Delta_B, \Delta_C)$ 是**正则**的乘子 Hopf 代数胚, 如果满足下列条件:

(i) 子空间 $t_C(_CI) \cdot A, I^C, A \cdot t_B(I_B), A \cdot {}^BI$ 都等于 A;

(ii) 典范映射 $T_\lambda, T_\rho, {}_\lambda T, {}_\rho T$ 都是双射.

我们称这样的乘子 Hopf 代数胚是**单位的**如果 A_B 和 A_C 是单位的.

接下来我们将说明上述典范映射是双射的等价于存在一个可逆的对极.

定义 5.4.11 一个乘子双代数胚 $\mathcal{A} = (A, B, C, t_B, t_C, \Delta_B, \Delta_C)$ 上的**可逆反对极**(invertible antipode) 指的是一个 A 到 A 的反自同构满足下列条件:

(i) S 到 $M(A)$ 的扩张满足 $S \circ t_B = \mathrm{id}_B, S \circ t_C = \mathrm{id}_C$;

(ii) 存在 A 上的一个左余单位 $_C\varepsilon$ 和右余单位 ε_B 满足

$$\varepsilon_B(a)b = S(a_{(1)})a_{(2)}b, \quad a_C\varepsilon(b) = ab_{(1)}S(b_{(2)})$$

对任意的 $a, b \in A$.

设 $\mathcal{A} = (A, B, C, t_B, t_C, \Delta_B, \Delta_C)$ 是一个乘子双代数胚, 记

$$\mathcal{A}^{\mathrm{co}} = (A, B, C, t_B^{-1}, t_C^{-1}, (\Delta_B)^{\mathrm{co}}, (\Delta_C)^{\mathrm{co}}),$$

$$\mathcal{A}^{\mathrm{op}} = (A, B, C, t_C^{-1}, t_B^{-1}, (\Delta_C)^{\mathrm{co}}, (\Delta_B)^{\mathrm{co}}),$$

$$\mathcal{A}^{\mathrm{op,co}} = (A^{\mathrm{op}}, C^{\mathrm{op}}, B^{\mathrm{op}}, t_C, t_B, (\Delta_C)^{\mathrm{op,co}}, (\Delta_B)^{\mathrm{op,co}}),$$

其中 $(\Delta_B)^{\mathrm{co}}, (\Delta_B)^{\mathrm{op}}, (\Delta_B)^{\mathrm{op,co}}$ 均是 Sweedler 记号.

读者可以直接证明如下的引理.

引理 5.4.12 设 $\mathcal{A}$ 是一个乘子双代数胚.

(i) 如果 $\mathcal{A}$ 是一个正则乘子 Hopf 代数胚, 则 $\mathcal{A}^{\mathrm{co}}, \mathcal{A}^{\mathrm{op}}$ 和 $\mathcal{A}^{\mathrm{op,co}}$ 也是正则的;

(ii) 如果 S 是 $\mathcal{A}$ 的一个可逆对极, 则 S 也是 $\mathcal{A}^{\mathrm{op,co}}$ 的可逆对极, 而 S^{-1} 为 $\mathcal{A}^{\mathrm{co}}$ 和 $\mathcal{A}^{\mathrm{op}}$ 的可逆对极.

下面我们就可以介绍本节的主要结论.

定理 5.4.13 设 $\mathcal{A}$ 是一个乘子双代数胚. 则 $\mathcal{A}$ 是正则的当且仅当存在 $\mathcal{A}$ 的可逆对极 S. 在此情形下, 可逆对极 S、左余单位 $_C\varepsilon$ 和右余单位 ε_B 都是唯一确定

的, 且下列等式成立:

$$\iota \otimes S = T_\rho \circ (\iota \otimes S) \circ {}_\rho T,$$
$$\iota \otimes S^{-1} = {}_\rho T \circ (\iota \otimes S^{-1}) \circ T_\rho.$$
$$S \otimes \iota = {}_\lambda T \circ (S \otimes \iota) \circ T_\lambda,$$
$$S^{-1} \otimes \iota = T_\lambda \circ (S^{-1} \otimes \iota) \circ {}_\lambda T.$$

证明 证明较为复杂, 本书略去, 可参考文献 [201].

接下来我们考虑乘子 Hopf 代数胚的例子.

例 5.4.14 从 [184] 可知, 给一个正则的弱乘子 Hopf 代数按照如下方式即可构造一个正则的乘子 Hopf 代数胚. 设 (A, Δ) 是一个正则的弱乘子 Hopf 代数, 那么存在一个正则的乘子 Hopf 代数胚 $(A, B, C, t_B, t_C, \Delta_B, \Delta_C)$ 使得

$$B = \varepsilon_s(A), \quad C = \varepsilon_t(A), \quad t_B = S^{-1} \circ \mathbf{id}_B, \quad t_C = S^{-1} \circ \mathbf{id}_C.$$

记

$$\pi_C : A \otimes A \to A^C \otimes {}_C A, \quad \pi_B : A \otimes A \to A_B \otimes {}^B A.$$

它们满足

$$\Delta_C(a)(1 \otimes b) = \pi_C(\Delta(a)(1 \otimes b)), \quad (a \otimes 1)\Delta_B(b) = \pi_B((a \otimes 1)\Delta(b))$$

对任意的 $a, b \in A$. 证明见 [184]. 在 [184] 中, 作者研究了什么时候一个正则的弱乘子 Hopf 代数会给出一个正则的乘子 Hopf 代数胚结构, 并给出了必要条件.

例 5.4.15 设 B 和 C 是两个非退化的幂等代数, 并且带有反同构 $S_B : B \to C$ 和 $S_C : C \to B$. 张量积 $A := C \otimes B$ 也是一个非退化的幂等代数. 注意到 B 和 C 到 $M(A)$ 之间有一个标准嵌入. 定义 $\Delta_C : A \to \text{End}(A^C \otimes {}_C A)$ 和 $\Delta_B : A \to \text{End}(A_B \otimes {}^B A)^{\text{op}}$ 为

$$\Delta_C(y \otimes x)(a \otimes a') = ya \otimes xa', \quad (a \otimes a')\Delta_B(y \otimes x) = ay \otimes a'x$$

对任意的 $a, a' \in A, x \in B, y \in C$. 那么 $\mathcal{A} = (A, B, C, S_C^{-1}, S_B^{-1}, \Delta_B, \Delta_C)$ 是一个正则的乘子 Hopf 代数胚. 其中余单位和对极定义如下:

$${}_C\varepsilon(y \otimes x) = yS_B(x), \quad \varepsilon_B(y \otimes x) = S_C(y)x, \quad S(y \otimes x) = S_B(x) \otimes S_C(y)$$

对任意的 $x \in B, y \in C$. 证明通过直接计算可得. 例如 [202] 中定理 5.9 的交换图可由下列两个等式推出:

$$(m_C \circ (S \otimes \mathbf{id}) \circ T_\rho)((y \otimes x) \otimes a) = S_C(y)xa = \varepsilon_B(y \otimes x)a,$$

$$(m_B \circ (\mathbf{id} \otimes S) \circ {}_\lambda T)(a \otimes (y \otimes x)) = ayS_B(x) = a_C\varepsilon(y \otimes x)$$

对任意的 $a \in A, x \in B, y \in C$. 如果在 $M(B \otimes C)$ 中存在一个正则的可分的幂等元,那么 A 上可以定义一个弱乘子 Hopf 代数结构, 见 [202]. 那么根据上面例子可知弱乘子 Hopf 代数 A 上可以构造一个正则的乘子 Hopf 代数胚, 并且我们知道正则的乘子 Hopf 代数胚与 $\mathcal{A}$ 同构.

例 5.4.16 设 B 和 C 是两个非退化的幂等代数, 并且带有反自同构 $S_B : B \to C$ 和 $S_C : C \to B$. 设 H 是一个正则的乘子 Hopf 代数并且在 C 上有一个单位的左作用和在 B 上有一个单位的右作用满足下列条件:

(i) B 和 C 都是 H 模代数.

(ii) 记 H 的对极为 S_H, 对任意的 $x \in B, y \in C, h \in H$,

$$S_B(x \triangleleft h) = S_H(h) \triangleright S_B(x) \quad \text{和} \quad S_C(h \triangleright y) = S_C(y) \triangleleft S_H(h).$$

那么空间 $A = C \otimes H \otimes B$ 是一个非退化的幂等的结合代数, 乘积定义为

$$(y \otimes h \otimes x)(y' \otimes h' \otimes x') = y(h_{(1)} \triangleright y') \otimes h_{(2)}h'_{(1)} \otimes (x \triangleleft h'_{(2)})x'.$$

代数 C, H, B 到 $M(A)$ 之间有自然嵌入, 那么有

$$yhx = y \otimes h \otimes x, \quad yxh = y \otimes h_{(1)} \otimes (x \triangleleft h_{(2)}), \quad hyx = (h_{(1)} \triangleright y) \otimes h_{(2)} \otimes x$$

对任意的 $h \in H, x \in B, y \in C$. 定义 $\Delta_C : A \to \mathrm{End}(A^C \otimes {}_C A)$ 和 $\Delta_B : A \to \mathrm{End}(A_B \otimes {}^B A)^{\mathrm{op}}$ 为

$$\Delta_C(yhx)(a \otimes a') = yh_{(1)}a \otimes h_{(2)}xa', \quad (a \otimes a')\Delta_B(yhx) = ayh_{(1)} \otimes a'h_{(2)}x$$

对任意的 $h \in H, x \in B, y \in C, a, a' \in A$. 那么 $\mathcal{A} = (A, B, C, S_C^{-1}, S_B^{-1}, \Delta_B, \Delta_C)$ 是一个正则的乘子 Hopf 代数胚. 其中余单位和对极定义如下:

$$_C\varepsilon(yxh) = yS_B(x)\varepsilon_H(h), \quad \varepsilon_B(hyx) = S_C(y)x\varepsilon_H(h), \quad S(yhx) = S_B(x)S_H(h)S_C(y)$$

对任意的 $x \in B, y \in C, h \in H$. 证明也是直接的. 同样地, 如果在 $M(B \otimes C)$ 中存在一个正则的可分的幂等元, 那么 A 上可以定义一个弱乘子 Hopf 代数结构, 见 [202]. 那么根据例 5.4.14 可知在弱乘子 Hopf 代数 A 上可以构造一个正则的乘子 Hopf 代数胚, 并且我们知道这个正则的乘子 Hopf 代数胚与 $\mathcal{A}$ 同构.

5.5 进一步研究的问题

这节主要给出一些目前国际前沿的研究问题或课题, 也是我们目前或今后进一步要讨论的理论 (参考 [239]—[249]). 这些理论的建立无疑在算子代数、泛函分析、数学物理、几何学、拓扑学和纯代数的交叉学科中起着非常重要的作用.

5.5.1 关于弱乘子 Hopf 代数问题

对于有限维弱 Hopf 代数来说,存在一个忠实的积分当且仅当代数是 Frobenius 的 (见 [25] 定理 3.16). 那么对于无限维弱乘子 Hopf 代数来说,情况变得更加复杂. 我们可以预测如果代数本身是 Frobenius 的并且存在足够多的积分的时候,那么肯定存在一个忠实的积分. 更进一步,在乘子 Hopf 代数理论中,积分是唯一的,因此存在足够多的积分一定可以导出存在一个忠实的积分. 那么在弱乘子 Hopf 代数情况下该如何考虑呢? 可以预测的是最终与基代数有关.

如果一个代数量子群胚存在一个忠实的积分,我们猜想它的对偶仍然存在一个忠实的积分,对于有限维弱 Hopf 代数来说,这个猜想是正确的. 但是,在弱乘子 Hopf 代数情况下,这个结论并不能由我们的命题 5.3.23 直接得到,还需要更深入的讨论.

对于一个弱乘子 Hopf 代数,如果存在一个忠实的积分,在 [203] 中作者证明了存在 Radon-Nikodym 元素,利用一个忠实的左 (右) 积分刻画了任意一个左 (右) 积分,证明了模元素的存在性,并给出了这个忠实的积分的模自同构. 如果没有假设存在一个忠实的积分,我们猜想上述结果仍然成立.

我们还可以考虑对合情形,也就是考虑弱乘子 Hopf ∗-代数. 在这种情况下,我们也可以得到平行的结果. 并且可以将整个结构提升到希尔伯特空间上去,这项工作目前正在进行中 [244].

对于弱乘子 Hopf 代数来说,我们只考虑了正则的情形. 有理由相信积分的存在性只能在正则的情况下研究,这是一个公开问题. 至今不是很清楚在非正则的情况下有哪些理论可以保留. 当然不仅是积分理论,更一般的理论也有待考虑. 对于乘子 Hopf 代数来说这也是一个公开的问题.

构造一个非正则的弱乘子 Hopf 代数的例子具有重要意义. Hopf 代数带有非双射的反对极显然是一个例子,但是我们应该考虑余乘不再是正则的以及反对极 S 不是从 A 到 A,因为弱 Hopf 代数是有单位的,所以不用考虑这种情况. 对于乘子 Hopf 代数来说,找一个非正则的乘子 Hopf 代数也是一个公开的问题.

5.5.2 关于乘子 Hopf 拟群问题

这方面研究已经完成 (见 [244]),这会涉及余乘子和余模余扩张理论. 例如,我们已经给出余乘子简单定义如下.

定义 5.5.1 设 (C, Δ) 是一个余代数. 余代数 C 的一个**左余乘子**是一个线性映射 $l: C \longrightarrow C$ 使得 $\Delta \circ l = (l \otimes 1) \circ \Delta$,即 $\sum l(c)_{(1)} \otimes l(c)_{(2)} = \sum l(c_{(1)}) \otimes c_{(2)}$ 对任意 $c \in C$,集合记为 $LC(C)$.

类似地,一个**右乘子**是一个线性映射 $r: C \longrightarrow C$ 使得 $\Delta \circ r = (1 \otimes r) \circ \Delta$,即 $\sum r(c)_{(1)} \otimes r(c)_{(2)} = \sum c_{(1)} \otimes r(c_{(2)})$ 对任意 $c \in C$,集合记为 $RC(C)$.

一个**余乘子**是左余乘子右余乘子对 (l,r) 使得 $(r \otimes 1) \circ \Delta = (1 \otimes l) \circ \Delta$, 即 $\sum r(c_{(1)}) \otimes c_{(2)} = \sum c_{(1)} \otimes l(c_{(2)})$ 对任意 $c \in C$, 集合记为 $CM(C)$.

例 5.5.2 设 $a^* \in C^*$, 定义线性映射 $l_{a^*} : C \longrightarrow C$ 为 $l_{a^*}(x) = \sum \langle a^*, x_{(1)} \rangle x_{(2)}$, 且定义 $r_{a^*} : C \longrightarrow C$ 为 $r_{a^*}(x) = \sum x_{(1)} \langle a^*, x_{(2)} \rangle$. 那么, $l_{a^*} \in LC(C), r_{a^*} \in RC(C)$ 和 $(l_{a^*}, r_{a^*}) \in CM$.

5.5.3 关于乘子余群胚问题

这方面研究已经完成 (见 [247]).

定义 5.5.3 设 I 是指标集, 且 $A = \{A_{ij} \mid i, j \in I\}$ 一族 $*$-代数 (可以没有单位元). 那么 $A = \{A_{ij} \mid i, j \in I\}$ 上的一个**余乘法**是一个 $*$-同态 $\Delta_{ij}^k : A_{ij} \longrightarrow M(A_{ik} \otimes A_{kj})$ 满足下面条件:

(MC1) $T_{1ij}^k(x \otimes y) = \Delta_{ij}^k(x)(1 \otimes y) \in A_{ik} \otimes A_{kj}$ 和 $T_{2ij}^k(a \otimes x) = (a \otimes 1)\Delta(x) \in A_{ik} \otimes A_{kj}$, 对任意 $x \in A_{ij}, y \in A_{kj}, a \in A_{ik}$;

(MC2) (**余结合性**) 对任意 $i, j, k, l \in I$, 在 $A_{ij} \otimes A_{jk} \otimes A_{kl}$ 有

$$(T_{2ik}^j \otimes \mathbf{id})(\mathbf{id} \otimes T_{1il}^k) = (\mathbf{id} \otimes T_{1jl}^k)(T_{2il}^j \otimes \mathbf{id}).$$

注记 条件 (MC1) 有意义, 因为我们有: $A_{ij} \otimes A_{kl} \subseteq M(A_{ij}) \otimes M(A_{kl}) \subseteq M(A_{ij} \otimes A_{kl})$.

定义 5.5.4 设 I 是指标集, 且 $A = \{A_{ij} \mid i, j \in I\}$ 一族 $*$-代数 (可以没有单位元). 设 $\Delta_{ij}^k : A_{ij} \longrightarrow M(A_{ik} \otimes A_{kj})$ 是 $A = \{A_{ij} \mid i, j \in I\}$ 上余乘法. 我们说 $A = \{A_{ij} \mid i, j \in I\}$ 是 I 上的一个**连通乘子余群胚**, 如果下面条件成立:

(MC3) (**连通性**) 所有 A_{ij} 是非零代数;

(MC4) 对任意 $x \in A_{ij}, y \in A_{kj}, a \in A_{ik}$, 线性映射 $T_{1ij}^k : A_{ij} \otimes A_{kj} \longrightarrow A_{ik} \otimes A_{kj}$, $x \otimes y \mapsto \Delta_{ij}^k(x)(1 \otimes y)$ 且 $T_{2ij}^k : A_{ik} \otimes A_{ij} \longrightarrow A_{ik} \otimes A_{kj}, a \otimes x \mapsto (a \otimes 1)\Delta_{ij}^k(x)$ 是双射.

引理 5.5.5 设 $A = \{A_{ij} \mid i, j \in I\}$ 是连通乘子余群胚. 那么存在一个同态 $\varepsilon_i : A_{ii} \longrightarrow \mathbb{C}$ 使得对任意 $i, j \in I$, 有

$$(\mathbf{id} \otimes \varepsilon_j)((c \otimes 1)\Delta_{ij}^j(x)) = cx = (\varepsilon_i \otimes \mathbf{id})(\Delta_{ij}^i(c)(1 \otimes x))$$

对任意 $c, x \in A_{ij}$.

引理 5.5.6 设 $A = \{A_{ij} \mid i, j \in I\}$ 是连通乘子余群胚. 那么存在一个反同态 $S_{ij} : A_{ij} \longrightarrow M(A_{ji})$ 使得

$$m_{ji}((1 \otimes S_{ij})((c \otimes 1)\Delta_{jj}^i(b))(1 \otimes d)) = c\varepsilon_j(b)d,$$
$$m_{ji}((c \otimes 1)(S_{ij} \otimes 1)(\Delta_{ii}^j(a)(1 \otimes d))) = c\varepsilon_i(a)d$$

对任意 $c,d \in A_{ji}, a \in A_{ii}, b \in A_{jj}$.

定理 5.5.7　如果 $A = \{A_{ij} \mid i,j \in I\}$ 是连通乘子余群胚带有单位元, 那么 A 是一个余群胚.

5.5.4　关于代数量子超群胚问题

这方面研究已经完成 (见 [242]).

定义 5.5.8　一个**弱乘子超代数**(weak multiplier hyperalgebra) A 定义为下面数据：

(1) 一个幂等结合代数具有非退化乘法 $\mu : A \otimes A \longrightarrow A$;

(2) 一个幂等元 $E \in M(A \otimes A)$;

(3) 一个线性映射 $\Delta : A \longrightarrow M(A \otimes A)$ (称为**余乘法**);

(4) 一个线性映射 $\varepsilon : A \longrightarrow k$ (称为**余单位**),

满足下面公理：

(i) 对任意 $a,b \in A$, $M(A \otimes A)$ 中的元素

$$T_1(a \otimes b) = \Delta(a)(1 \otimes b) \quad \text{和} \quad T_2(a \otimes b) = (a \otimes 1)\Delta(b)$$

在 $A \otimes A$ 的双边理想中.

(ii) 余乘法是结合的, 即满足

$$(T_2 \otimes \mathbf{id})(\mathbf{id} \otimes T_1) = (\mathbf{id} \otimes T_1)(T_2 \otimes \mathbf{id}).$$

(iii) 余单位满足：

$$(\varepsilon \otimes \mathbf{id})T_1 = \mu = (\mathbf{id} \otimes \varepsilon)T_2.$$

(iv) 关于元 E 和 Δ, 有

$$\langle \Delta(a)(b \otimes c) \mid a,b,c \in A \rangle = \langle E(b \otimes c) \mid b,c \in A \rangle$$

和

$$\langle (b \otimes c)\Delta(a) \mid a,b,c \in A \rangle = \langle (b \otimes c)E \mid b,c \in A \rangle.$$

(v) 在 $M(A \otimes A \otimes A)$ 中, 幂等元 E 满足

$$(E \otimes 1)(1 \otimes E) = E^{(3)} = (1 \otimes E)(E \otimes 1).$$

(vi) 对任意 $a,b,c \in A$, 有

$$(\varepsilon \otimes \mathbf{id})((1 \otimes a)E(b \otimes c)) = (\varepsilon \otimes \mathbf{id})(\Delta(a)(b \otimes c))$$

和
$$(\varepsilon \otimes \mathbf{id})((a \otimes b)E(1 \otimes c)) = (\varepsilon \otimes \mathbf{id})((a \otimes b)\Delta(c)).$$

容易得到: $E(a) = (a) = (a)E$ 对任意 $a \in A$. 另外, 可定义

$$a\varepsilon_s(b) - (\mathbf{id} \otimes \varepsilon)((a \otimes b)E) \quad 和 \quad \varepsilon_t(a)b - (\varepsilon \otimes \mathbf{id})(E(a \otimes b)).$$

和
$$\widetilde{\varepsilon}_s(b)a = (\mathbf{id} \otimes \varepsilon)(E(a \otimes b)) \quad 和 \quad b\widetilde{\varepsilon}_t(a) = (\varepsilon \otimes \mathbf{id})((a \otimes b)E).$$

定义 5.5.9 我们说 A 是**正则的**, 如果 $T_3(a \otimes b) = \Delta(a)(b \otimes 1)$ 和 $T_4(a \otimes b) = (1 \otimes a)\Delta(b)$ 都在 $A \otimes A$ 中, 对任意 $a, b \in A$.

定义 5.5.10 设 $(A, \Delta, \varepsilon, E)$ 是一个弱乘子超代数, 那么

(i) 一个线性函数 $\varphi : A \longrightarrow \mathbb{C}$ 称为是**左不变量** (left invariant) 如果 $(1 \otimes \varphi)\Delta(a) \in A_t$ 对任意 $a \in A$. 一个非零的左不变量函数称为 A 上的**左积分** (left integral);

(ii) 类似地, 一个线性函数 $\psi : A \longrightarrow \mathbb{C}$ 称为是**右不变量** (right invariant) 如果 $(\psi \otimes 1)\Delta(1) \in A_s$ 对任意 $a \in A$. 一个非零的右不变量函数称为 A 上的**右积分** (right integral).

定义 5.5.11 设 $(A, \Delta, \varepsilon, E)$ 是一个弱乘子超代数具有忠实左积分 φ. 假设存在一个线性双射 $S : A \longrightarrow A$ 满足:

$$S((\mathbf{id} \otimes \varphi)T_1(a \otimes b)) = (\mathbf{id} \otimes \varphi)T_4(a \otimes b)$$

对任意 $a, b \in A$. 如果进一步 S 是反同态, 那么 S 称为关于 φ 的**反对极**.

命题 5.5.12 定义 $\psi = \varphi \circ S$. 那么 ψ 是 A 上的忠实右积分. 进一步, 有

$$S((\psi \otimes \mathbf{id})T_2(a \otimes b)) = (\psi \otimes \mathbf{id})T_3(a \otimes b)$$

对任意 $a, b \in A$.

定义 5.5.13 设 $(A, \Delta, \varepsilon, E)$ 是一个弱乘子超代数具有一个正则余乘法. 假设存在一个忠实左积分 φ 和相关的反对极 S. 那么 $(A, \Delta, \varepsilon, E, \varphi, S)$ 称为**代数量子超群胚** (algebraic quantum hypergroupoid). 进一步, 如果 A 是 $*$-代数且 Δ 是 $*$- 映射, 那么称 $(A, \Delta, \varepsilon, E)$ 是 $*$-代数量子超群胚.

例 5.5.14 (i) 一个代数量子群胚 [57] 是一个代数量子超群胚;

(ii) 一个有限维弱 Hopf 代数 [25] 是一个代数量子超群胚;

(iii) 一个广义弱双 Frobenius 代数 [41] 是一个代数量子超群胚;

(iv) 一个双 Frobenius 代数 [64] 是一个代数量子超群, 所以它是一个代数量子超群胚;

(v) 一个超群胚上的代数函数具有有限支撑, 可以构成一个代数量子超群胚[242].

接下来主要研究该对象的 Pontryagin 对偶问题和其他相关问题, 例如: 有界型上的理论建立.

5.5.5 其他问题

在第 4 章中, 我们介绍了一些 Hopf 型代数结构, 问题是: 如果考虑这些代数的乘子结构, 如何进行研究? 因此就有大量问题摆在我们面前需要讨论. 根据我们的经验, 研究这些问题不是那么容易的, 需要学习算子代数的一些理论和掌握一些基本例子, 从而产生灵感和新思想方法.

第6章 非线性方程与微积分

本章主要介绍 Hopf 代数中的一些非线性方程与微积分理论, 展示出如何通过这些方程来研究某类 Hopf 代数相关的范畴, 反之亦然. 当然这些思想方法可以用于第 4 章中介绍的各类 Hopf 型代数中, 因此有大量问题可进行研究.

6.1 非线性方程

本节主要介绍一些与量子群相关的非线性方程或方程组, 这些方程与 Hopf 代数表示和余表示范畴的研究将在 6.2 节讨论.

6.1.1 量子杨–巴斯特方程

首先介绍杨–巴斯特方程 [189, 256], 该问题来源于统计力学与低维拓扑理论.

定义 6.1.1 (i) 设 V 是向量空间, $R: V \otimes V \longrightarrow V \otimes V$ 是一个线性映射. 下面等式称为**量子杨–巴斯特**方程 (quantum Yang-Baxter equation(QYBE)):

$$R^{12}R^{13}R^{23} = R^{23}R^{13}R^{12}, \tag{6.1.1}$$

这里 $R^{12} = R \otimes \mathbf{id}_V, R^{23} = \mathbf{id}_V \otimes R, R^{13} = (\tau \otimes \mathbf{id}_V)R^{23} = (\mathbf{id}_V \otimes \tau)R^{12}$ 等.

(ii) 设 V 是向量空间, $c: V \otimes V \longrightarrow V \otimes V$ 是一个线性映射. 下面等式称为**辫子方程** (braided equation):

$$c^{12}c^{23}c^{12} = c^{23}c^{12}c^{23},$$

即

$$(c \otimes \mathbf{id})(\mathbf{id} \otimes c)(c \otimes \mathbf{id}) = (\mathbf{id} \otimes c)(c \otimes \mathbf{id})(\mathbf{id} \otimes c).$$

由此, 不难看出, QYBE 与辫子方程的解是一一对应的.

注记 菲尔兹奖得主、苏联数学物理学家 V.Drinfel'd 教授指出, 如果 V 仅是一个集合, $R: V \times V \longrightarrow V \times V$ 仅是集合映射, 同样有 QYBE 与辫子方程的解是一一对应的. 他称这个是**集合理论量子杨–巴斯特方程**, 并且认为是一个有意义的问题. 集合理论量子杨–巴斯特方程的解可以线性扩张成向量范畴中的量子杨–巴斯特方程的解.

例 6.1.2 (i) 我们有: R 是辫子方程的解, 当且仅当 $R\tau$ 是量子杨–巴斯特方程的解, 当且仅当 τR 是量子杨–巴斯特方程的解, 当且仅当 $R\tau R$ 是辫子方程的解.

(ii) 设 V 是向量空间，设 $\{k_{ij} \in k \mid i,j = 1,2,\cdots,n\}$，定义

$$R: V \otimes V \longrightarrow V \otimes V, \quad v_i \otimes v_j \mapsto k_{ij} v_j \otimes v_i,$$

那么 R 是量子杨–巴斯特方程的解. 尤其，$\mathbf{id}_{V \otimes V}$ 和扭曲映射 τ 是量子杨–巴斯特方程和辫子方程的解.

(iii) 设 V 是有限维向量空间，u 是 V 的自同构. 如果 R 是量子杨–巴斯特方程的解，那么 $T = (u \otimes u) R (u \otimes u)^{-1}$ 也是量子杨–巴斯特方程的解.

(iv) R 是量子杨–巴斯特方程的解，当且仅当 R^{-1} 是量子杨–巴斯特方程的解.

(v) 设 V 是向量空间，$f, g \in \mathrm{End}_k(V)$ 满足 $gf = fg$，那么 $R = f \otimes g$ 是量子杨–巴斯特方程的解.

例如：设 $\dim(V) = 2$，且 $\{e_1, e_2\}$ 是 V 的一组基，那么 $f, g \in \mathrm{End}_k(V)$ 在这组基下的矩阵分别为

$$f = \begin{pmatrix} a & 1 \\ 0 & a \end{pmatrix}, \qquad g = \begin{pmatrix} b & c \\ 0 & b \end{pmatrix}$$

对任意 $a, b, c \in k$. 那么，在基 $\{e_1 \otimes e_1, e_1 \otimes e_2, e_2 \otimes e_1, e_2 \otimes e_2\}$ 下，$R = f \otimes g$ 的矩阵为

$$f = \begin{pmatrix} ab & ac & b & c \\ 0 & ab & 0 & b \\ 0 & 0 & ab & ac \\ 0 & 0 & 0 & ab \end{pmatrix},$$

因为 $fg = gf$，所以 $R = f \otimes g$ 是量子杨–巴斯特方程的解.

(vi) 设 $0 \neq q \in k$，V 是有限维向量空间. 设 $R_q : V \otimes V \longrightarrow V \otimes V$ 为

$$R_q(v_i \otimes v_j) = \begin{cases} q v_i \otimes v_i, & v_i = v_j, \\ v_i \otimes v_j, & i \leqslant j, \\ v_i \otimes v_j + (q - q^{-1}) v_j \otimes v_i, & i > j, \end{cases}$$

所以 R 是量子杨–巴斯特方程的解.

(vii) 设 A 是一个代数. 考虑映射

$$R: B \otimes B \longrightarrow B \otimes B, a \otimes b = ab \otimes 1 + 1 \otimes ab - a \otimes b,$$

那么容易验证，R 是辫子方程的解.

对偶地，设 C 是一个余代数. 考虑映射

$$R: C \otimes C \longrightarrow C \otimes C, c \otimes d = \varepsilon(c) \sum d_1 \otimes d_2 + \varepsilon(d) \sum c_1 \otimes c_2 - d \otimes c,$$

那么容易验证, R 是量子杨–巴斯特方程的解.

(viii) 设 S 是一个集合且 $\triangleright : S \otimes S \longrightarrow S, (x,y) \mapsto x \triangleright y$ 是一个映射. 那么 $R : S \otimes S \longrightarrow S \otimes S, (x,y) \mapsto (x \triangleright y, y)$ 是辫子方程的解当且仅当

$$x \triangleright (y \triangleright z) = (x \triangleright y) \triangleright (x \triangleright z)$$

对任意 $x,y,z \in S$.

如果上面等式成立且对任意 $x \in S$, 映射 $s\triangleright : S \longrightarrow S, y \mapsto x \triangleright y$ 是双射, 那么称 $(S, \triangleright)$ 为**架**(rack). 此概念可用于构造纽结与链不变量.

设 $\mathfrak{h}$ 是有限维交换李代数, V 是半单 $\mathfrak{h}$-模. 记 M 是 $\mathfrak{h}$ 上亚纯 (meromorphic) 函数域, M 有一个平凡 $\mathfrak{h}$-模结构.

定义 6.1.3 设 $R : V \otimes V \otimes M \longrightarrow V \otimes V \otimes M$ 是一个 $\mathfrak{h}$-不变的 M-线性映射. 那么关于 R 的**量子动力杨–巴斯特方程**(QDYBE) 是下面等式:

$$R^{12}(\lambda - h^{(3)}) R^{13}(\lambda) R^{23}(\lambda - h^{(1)}) = R^{23}(\lambda) R^{13}(\lambda - h^{(2)}) R^{12}(\lambda) \tag{6.1.2}$$

对任意 $\lambda \in \mathfrak{g}^*, h \in \mathfrak{h}$.

一个**量子动力 R-矩阵**是这个方程的一个可逆解.

注记 6.1.4 (i) 当 $\mathfrak{h} = 0$ 时, 方程 (6.1.2) 就变成一般量子杨–巴斯特方程 (6.1.1).

(ii) 量子动力杨–巴斯特方程 (6.1.2) 的常量解与普通量子杨–巴斯特方程 (6.1.1) 的 $\mathfrak{h}$-不变解是一样的.

(iii) 在物理文献中, 变量 λ 称为是**动力变量**. 这自然产生 "**动力 R-矩阵**" 的概念. 在方程 (6.1.2) 中, 令 $\lambda := \dfrac{\lambda}{\gamma}$, 则有

$$\widetilde{R}^{12}(\lambda - \gamma h^{(3)}) \widetilde{R}^{13}(\lambda) \widetilde{R}^{23}(\lambda - \gamma h^{(1)}) = \widetilde{R}^{23}(\lambda) \widetilde{R}^{13}(\lambda - \gamma h^{(2)}) \widetilde{R}^{12}(\lambda). \tag{6.1.3}$$

我们称 (6.1.3) 为带有步的动力杨–巴斯特方程.

更一般地, 设 $\mathfrak{g}$ 是一个单复李代数, $\mathfrak{h} \subset \mathfrak{g}$ 是 Cartan 子代数且 $\Delta \subset \mathfrak{h}^*$ 是结合的根系, 设 Π 是单根集合, $\Delta^+ \subset \Delta$ 是结合的正根系. 设 $\mathfrak{g} = \mathfrak{n}_- \oplus \mathfrak{h} \oplus \mathfrak{n}_+$ 是对应的三角分解 (polarization), $\mathfrak{g}_\alpha$ 是 $\mathfrak{g}$ 的根子空间. 设 $\langle , \rangle$ 是 $\mathfrak{g}$ 上的非退化不变对称形, 对于长根其正规化为 $\langle \alpha, \alpha \rangle = 2$. 最后, 对任意 $\alpha \in \Delta$, 选择一个 $e_\alpha \in \mathfrak{g}_\alpha$ 使得 $\langle e_\alpha, e_{-\alpha} \rangle = 1$.

对 $\lambda \in \mathfrak{h}^*$, 设 $\mathbb{C}_\lambda$ 是一维 $(\mathfrak{h} \oplus \mathfrak{n}_+)$-模使得 $\mathbb{C}_\lambda = \mathbb{C} x_\lambda$ 具有 $h \cdot x_\lambda = \lambda(h) x_\lambda$ 对于任意 $h \in \mathfrak{h}$ 和 $\mathfrak{n}_+ \cdot x_\lambda = 0$. 一个最高权 (highest weight) 的 Verma 模是诱导模:

$$M_\lambda = \mathrm{Ind}_{\mathfrak{h} \oplus \mathfrak{n}_+}^{\mathfrak{g}} \mathbb{C}_\lambda.$$

注意: M_λ 是一个自由 $U(\mathfrak{n}_-)$-模且作为线性空间等同于 $U(\mathfrak{n}_-)$: $U(\mathfrak{n}_-) \cong M_\lambda, u \mapsto u \cdot x_\lambda$.

在 $\mathfrak{h}^*$ 上可以定义一个偏序: $\mu < \nu$ 如果存在 $\alpha_1, \cdots, \alpha_r \in \Delta^+, r > 0$, 使得 $\nu = \mu + \alpha_1 + \cdots + \alpha_r$. 设 $M_\lambda = \bigoplus_{\mu \leq \nu} M_\lambda[\mu]$ 是 M_λ 的权子空间分解.

命题 6.1.5 对 λ 的泛 (generic) 数值, 模 M_λ 是不可约的.

也可以定义对偶 Verma 模 M_λ^*, 它是一个分次对偶空间 $\bigoplus_\mu M_\lambda[\mu]^*$, 具有下面 $\mathfrak{g}$- 作用:

$$(a \cdot u)(v) = -u(a \cdot v), \quad \forall a \in \mathfrak{g}, u \in M_\lambda^*, v \in M_\lambda.$$

设 x_λ^* 是 M_λ^* 的最低权向量使得 $\langle x_\lambda, x_\lambda^* \rangle = 1$.

现在设 V 是一个有限维 $\mathfrak{g}$-模. 令 $V = \bigoplus_{\nu \in \mathfrak{h}^*} V[\nu]$ 是它的权子空间分解. 设 $\lambda, \mu \in \mathfrak{h}^*$, 我们考虑 $\mathfrak{g}$-模扭曲 (intertwining) 算子

$$\Phi: M_\lambda \longrightarrow M_\mu \otimes V.$$

进而, 可以定义**期望值**(expectation value) 为

$$\langle \Phi \rangle = \langle \Phi \cdot x_\lambda, x_\mu^* \rangle \in V[\lambda - \mu],$$

此定义相似于量子场论中的期望值.

命题 6.1.6 设 M_μ 不可约, 那么映射

$$\mathrm{Hom}_\mathfrak{g}(M_\lambda, M_\mu \otimes V) \longrightarrow V[\lambda - \mu], \quad \Phi \mapsto \langle \Phi \rangle$$

是一个同构.

证明 由 Frobenius 相互作用 (reciprocity), 有

$$\mathrm{Hom}_\mathfrak{g}(M_\lambda, M_\mu \otimes V) = \mathrm{Hom}_{\mathfrak{h} \oplus \mathfrak{n}_+}(\mathbb{C}_\lambda, M_\mu \otimes V) = \mathrm{Hom}_{\mathfrak{h} \oplus \mathfrak{n}_+}(\mathbb{C}_\lambda \otimes M_\mu^*, V).$$

进一步, 因为 M_μ 是不可约的, 所以有 $M_\mu^* = \mathrm{Ind}_\mathfrak{h}^{\mathfrak{h} \oplus \mathfrak{n}_+} \mathbb{C}_{-\mu}$ 作为 $\mathfrak{h} \oplus \mathfrak{n}_+$-模. 尤其, 有

$$\mathrm{Hom}_{\mathfrak{h} \oplus \mathfrak{n}_+}(\mathbb{C}_\lambda \otimes M_\mu^*, V) = \mathrm{Hom}_\mathfrak{h}(\mathbb{C}_\lambda \otimes \mathbb{C}_{-\mu}, V) = V[\lambda - \mu].$$

该命题可以叙述为: 对任意 $v \in V[\lambda - \mu]$ 存在唯一扭曲算子 $\Phi_\lambda^v : M_\lambda \longrightarrow M_\mu \otimes V$ 使得

$$\Phi_\lambda^v(x_\lambda) \in x_\mu \otimes v + \bigoplus_{\nu < \mu} M_\mu[\nu] \otimes V.$$

注意: 对固定 v, Φ_λ^v 只定义在 λ 的泛数值上. 把 Verma 模 M_λ, M_μ 与 $U(\mathfrak{n}_-)$ 等同起来. 我们有线性映射 $\Phi_\lambda^v : U(\mathfrak{n}_-) \longrightarrow U(\mathfrak{n}_-) \otimes V$. 容易看出, 对任意基, 这个映射的系数是 λ 的有理函数.

6.1 非线性方程

设 $wt(u) \in \mathfrak{h}^*$ 表示任意齐次向量 u 在 $\mathfrak{g}$-模中的权. 设 V, W 是两个有限维 $\mathfrak{g}$-模, 且 $v \in V, w \in W$ 是两个齐次向量. 设 $\lambda \in \mathfrak{h}^*$. 考虑下面合成映射:

$$\Phi_\lambda^{w,v}: M_\lambda \xrightarrow{\Phi_\lambda^v} M_{\lambda-wt(v)} \otimes V \xrightarrow{\Phi_{\lambda-wt(v)}^w} M_{\lambda-wt(v)-wt(w)} \otimes W \otimes V,$$

这里 Φ 代表 $\Phi \otimes 1$, 以下类似. 那么, 我们有 $\Phi_\lambda^{w,v} \in \operatorname{Hom}_{\mathfrak{g}}(M_\lambda, M_{\lambda-wt(v)-wt(w)} \otimes W \otimes V)$.

于是, 由上面命题, 对任意泛数值 λ, 存在唯一元素 $u \in W + V[wt(v) + wt(w)]$ 使得 $\Phi_\lambda^u = \Phi_\lambda^{w,v}$. 显然对应 $(w, v) \mapsto u$ 是双线性的, 且定义了一个 $\mathfrak{h}$-模线性映射:

$$J_{WV}(\lambda): W \otimes V \longrightarrow W \otimes V, \quad w \otimes v \mapsto \langle \Phi_\lambda^{w,v} \rangle.$$

我们称算子 $J_{WV}(\lambda)$ 是 V 和 W 的**聚矩阵**(fusion matrix).

关于聚矩阵, 我们有下面基本性质.

如果 $A_1, \cdots, A_r$ 是半单 $\mathfrak{h}$-模, 且 $F(\lambda): A_1 \otimes \cdots \otimes A_r \longrightarrow A_1 \otimes \cdots \otimes A_r$ 是依赖于 $\lambda \in \mathfrak{h}^*$ 的线性算子. 那么, 对于任意齐次元素 $a_1, \cdots, a_r$, 设

$$F(\lambda - h^{(i)})(a_1 \otimes \cdots \otimes a_r) := F(\lambda - wt(a_i))(a_1 \otimes \cdots \otimes a_r).$$

命题 6.1.7 设 V, W 是两个有限维 $\mathfrak{g}$-模, 那么

(i) $J_{WV}(\lambda)$ 是 λ 的有理函数.

(ii) $J_{WV}(\lambda)$ 是严格下三角矩阵, 即 $J = 1 + N$, 这里

$$N(W[\nu] \otimes V[\mu]) \subset \bigoplus_{\tau < \nu, \mu < \sigma} W[\tau] \otimes V[\sigma].$$

尤其, J_{WV} 是可逆的.

(iii) 设 U, V, W 是三个有限维 $\mathfrak{g}$-模, 那么在 $U \otimes W \otimes V$ 上, 聚矩阵满足下面动力 2-余循环条件:

$$J_{U \otimes W, V}(\lambda)(J_{UW}(\lambda - h^{(3)}) \otimes 1) = J_{U, W \otimes V}(\lambda)(1 \otimes J_{WV}(\lambda)).$$

设 V, W 是两个有限维 $\mathfrak{g}$-模. 定义

$$R_{VW}(\lambda) = J_{VW}(\lambda)^{-1} J_{WV}^{21}(\lambda) \in \operatorname{Hom}_{\mathfrak{h}}(V \otimes W, V \otimes W),$$

这里 $J^{21} = PJP$ 具有 $P(x \otimes y) = y \otimes x$, 即 $R_{VW}(\lambda)(v \otimes w) = \sum_i v_i \otimes w_i$, 这里 $\Phi_\lambda^{w,v} = P \sum_i \Phi_\lambda^{w_i, v_i}$. 算子 $R_{VW}(\lambda)$ 称为 V 和 W 的**变换矩阵**(exchange matrix).

命题 6.1.8 设 U, V, W 是三个有限维 $\mathfrak{g}$-模. 那么在代数 $\operatorname{Hom}_{\mathfrak{h}}(V \otimes W \otimes U, V \otimes W \otimes U)$ 中, 变换矩阵满足下面关系式:

$$R_{VW}(\lambda - h^{(3)}) R_{VU}(\lambda) R_{WU}(\lambda - h^{(1)}) = R_{WU}(\lambda) R_{VU}(\lambda - h^{(2)}) R_{VW}(\lambda).$$

例 6.1.9 我们看一个最简单的聚矩阵与变换矩阵的例子. 取 $\mathfrak{g} = \mathfrak{sl}_2 = \mathbb{C}e \oplus \mathbb{C}h \oplus \mathbb{C}f$ 和 $V = \mathbb{C}^2 = \mathbb{C}v_+ \oplus \mathbb{C}v_-$ 满足

$$h \cdot v_\pm = \pm v_\pm, \quad e \cdot v_- = v_+, \quad e \cdot v_+ = 0, \quad f \cdot v_- = 0, \quad f \cdot v_+ = v_-.$$

我们先计算聚矩阵 $J_{VV}(\lambda)$. 由 $J_{VV}(\lambda)$ 的三角性质, 有

$$J_{VV}(\lambda)(v_\pm \otimes v_\pm) = v_\pm \otimes v_\pm, \quad J_{VV}(\lambda)(v_- \otimes v_+) = v_- \otimes v_+,$$

所以, 只剩下计算 $J_{VV}(\lambda)(v_+ \otimes v_-)$. 考虑扭曲子 $\Phi_\lambda^{v_-} : M_\lambda \longrightarrow M_{\lambda+1} \otimes V$, 由定义可知, $\Phi_\lambda^{v_-}(x_\lambda) = x_{\lambda+1} \otimes v_- + y(\lambda) f x_{\lambda+1} \otimes v_+$. 为了决定 $y(\lambda)$ 函数, 用扭曲性质:

$$0 = \Phi_\lambda^{v_-}(e x_\lambda) = (e \otimes 1 + 1 \otimes e)\Phi_\lambda^{v_-}$$
$$= x_{\lambda+1} \otimes v_+ + y(\lambda) e f x_{\lambda+1} \otimes v_+$$
$$= x_{\lambda+1} \otimes v_+ + y(\lambda)(h + fe) x_{\lambda+1} \otimes v_+$$
$$= x_{\lambda+1} \otimes v_+ + (\lambda+1) y(\lambda) x_{\lambda+1} \otimes v_+.$$

因此, $y(\lambda) = -\dfrac{1}{\lambda+1}$. 也显然有: $\Phi_{\lambda+1}^{v_+}(x_{\lambda+1}) = x_\lambda \otimes v_+$. 所以,

$$\Phi_\lambda^{v_+, v_-}(x_\lambda) = \Phi_{\lambda+1}^{v_+} \Phi_\lambda^{v_-}(x_\lambda) = x_\lambda \otimes \left(v_+ \otimes v_- - \frac{1}{\lambda+1} v_- \otimes v_+ \right) + \text{低权项}.$$

因此, $J_{VV}(\lambda)(v_+ \otimes v_-) = v_+ \otimes v_- - \dfrac{1}{\lambda+1} v_- \otimes v_+$, 且

$$J_{VV}(\lambda) = \begin{pmatrix} 1 & 0 & 0 & 0 \\ 0 & 1 & 0 & 0 \\ 0 & -\dfrac{1}{\lambda+1} & 1 & 0 \\ 0 & 0 & 0 & 1 \end{pmatrix},$$

在基 $(v_+ \otimes v_+, v_+ \otimes v_-, v_- \otimes v_+, v_- \otimes v_-)$ 上的, 变换矩阵为

$$R_{VV}(\lambda) = \begin{pmatrix} 1 & 0 & 0 & 0 \\ 0 & 1 & -\dfrac{1}{\lambda+1} & 0 \\ 0 & \dfrac{1}{\lambda+1} & 1 - \dfrac{1}{(\lambda+1)^2} & 0 \\ 0 & 0 & 0 & 1 \end{pmatrix}.$$

6.1.2 Hopf 方程 (参考文献 [34])

定义 6.1.10 (i) 设 V 是向量空间, $R \in \mathrm{End}_k(V \otimes V)$. 我们说 R 是**五角** (pentagon) **方程**的解, 如果

$$R^{12}R^{13}R^{23} = R^{23}R^{12}.$$

(ii) 我们说 R 是 **Hopf 方程**的解, 如果

$$R^{23}R^{13}R^{12} = R^{12}R^{23}.$$

命题 6.1.11 设 V 是向量空间, $R \in \mathrm{End}_k(V \otimes V)$. 下面叙述等价:
(i) R 是 Hopf 方程的解;
(ii) $T = \tau R$ 是方程: $T^{12}T^{23}T^{12} = T^{23}\tau^{12}T^{23}$ 的解;
(iii) $T = R\tau$ 是方程: $T^{23}T^{12}T^{23} = T^{12}T^{13}\tau^{23}$ 的解;
(iv) $W = \tau R\tau$ 是五角方程的解.

定义 6.1.12 设 V 是向量空间, 如果 $R \in \mathrm{End}_k(V \otimes V)$ 是 Hopf 方程的解.
(i) 我们说 R 是交换的, 如果 $R^{12}R^{13} = R^{13}R^{12}$;
(ii) 我们说 R 是余交换的, 如果 $R^{13}R^{23} = R^{23}R^{13}$.

例 6.1.13 (i) 设 V 是向量空间, 那么 $\mathrm{id}_{V \otimes V} \in \mathrm{End}_k(V \otimes V)$ 是 Hopf 方程的解.

(ii) 设 V 是有限维向量空间, u 是 V 的自同构. 如果 R 是 Hopf 方程的解, 那么 $T = (u \otimes u)R(u \otimes u)^{-1}$ 也是 Hopf 方程的解.

(iii) 设 V 是向量空间, $f, g \in \mathrm{End}_k(V)$ 满足 $f^2 = f, g^2 = g, gf = fg$, 那么 $R = f \otimes g$ 是 Hopf 方程的解.

(iv) 设 B 是一个双代数. 考虑映射 $T_2 : B \otimes B \longrightarrow B \otimes B$, $a \otimes b = (a \otimes 1)\Delta(b)$, 那么容易验证, T_2 是 Hopf 方程的解.

(v) 设 B 是一个双代数. 考虑映射 $T_1 : B \otimes B \longrightarrow B \otimes B$, $a \otimes b = \Delta(a)(1 \otimes b)$, 那么容易验证, T_1 是五角方程的解.

(vi) 设 A 是一个代数, $a, b \in A$. 令 $R = 1 \otimes 1 + a \otimes b \in A \otimes A$. 那么, R 是五角方程的解当且仅当

$$a \otimes (ab - ba - 1) \otimes b = aoa \otimes b^2 + a^2 \otimes b \otimes b + a^2 \otimes ba \otimes b^2. \tag{6.1.4}$$

进一步地, 如果 a, b 是幂零元素, 那么 R 是可逆的. 现在取 a, b 满足

$$\begin{cases} a^2 = b^2 = 0, \\ ab - ba = 1, \end{cases} \tag{6.1.5}$$

或
$$\begin{cases} a^2 = 0, \\ b^2 = bab - ba = a + 1, \end{cases} \tag{6.1.6}$$

那么, (6.1.4) 成立, 所以 R 是五角方程的解.

我们现在给出 $A = M_n(k)$ 使得 $\mathrm{char}(k)|n$ 成立的一些详细例子. 假设 $\mathrm{char}(k) = 2$ 且 $n = 2q$, 这里 q 是一个正整数. 那么

$$a = e_1^2 + e_3^4 + \cdots + e_{2q-1}^{2q}, \quad b = e_2^1 + e_4^3 + \cdots + e_{2q}^{2q-1}$$

满足 (6.1.5), 因此

$$R = I_n \otimes I_n + \sum_{i,j=1}^{q} e_{2i-1}^{2i} \otimes e_{2j}^{2j-1}$$

是五角方程的可逆解.

另一方面, 对任意一个可逆矩阵 $X \in M_q(k)$, 如下给出的矩阵 $a, b \in M_n(k)$

$$a = \begin{pmatrix} I_q & X^{-1} \\ X & I_q \end{pmatrix}, \quad b = e_1^1 + e_2^2 + \cdots + e_q^q$$

满足 (6.1.6), 因此

$$R = R_X = I_n \otimes I_n + \sum_{i=1}^{q} a \otimes e_i^i$$

是五角方程的可逆解.

6.1.3 Long 方程

定义 6.1.14　设 V 是向量空间, $R \in \mathrm{End}_k(V \otimes V)$. 我们说 R 是 **Long 方程** 的解, 如果

$$R^{12}R^{23} = R^{23}R^{12}.$$

注记　我们称 $R^{12}R^{23} = R^{23}R^{12} = R^{13}R^{12}$ 为 **FS-方程**(Frobenius-separability equation).

例 6.1.15　(i) 设 V 是向量空间且 $R \in \mathrm{End}_k(V \otimes V)$. 那么 R 是 Long 方程的解当且仅当 R^{-1} 是 Long 方程的解.

(ii) 设 V 是向量空间, 设 $\{k_{ij} \in k \mid i,j = 1, 2, \cdots, n\}$, 定义

$$R : V \otimes V \longrightarrow V \otimes V, \quad v_i \otimes v_j \mapsto k_{ij} v_j \otimes v_i,$$

那么 R 是 Long 方程的解. 尤其, $\mathrm{id}_{V \otimes V}$ 是 Long 方程的解.

(iii) 设 V 是有限维向量空间, u 是 V 的自同构. 如果 R 是 Long 方程的解, 那么 $T = (u \otimes u)R(u \otimes u)^{-1}$ 也是 Long 方程的解.

命题 6.1.16 设 V 是向量空间, $R \in \mathrm{End}_k(V \otimes V)$. 下面叙述等价:

(i) R 是 Long 方程的解;

(ii) $T = R\tau$ 是方程: $T^{12}T^{13} = T^{23}T^{13}\tau^{(123)}$ 的解;

(iii) $U = \tau R$ 是方程: $U^{13}T^{23} = \tau^{(123)}U^{13}U^{12}$ 的解;

(iv) $W = \tau R\tau$ 是方程: $W^{12}W^{13}\tau^{(123)} = \tau^{(123)}W^{23}W^{13}$ 的解.

注记 如果 R 是 Long 方程的解, 那么 R 满足 Knizhnik-Zamolodchikov 方程可积性条件:
$$[R^{12}, R^{13} + R^{23}] = 0.$$

6.1.4 微积分

定义 6.1.17 设 A 是一个带有单位元的代数. Γ 是 A 上的双模且 $d: A \longrightarrow \Gamma$ 是一个线性映射. 我们说 (Γ, d) 是 A 上的**一阶微积分**(the first order differential calculus) 如果下面条件成立:

(i) **Leibniz 法则** 对任意 $a, b \in A$, 有 $d(ab) = d(a)b + ad(b)$;

(ii) **标准形式**(standard form) 任意元素 $q \in \Gamma$ 具有形式: $q = \sum_i^n a_i \mathrm{d}(b_i)$, 这里 $a_i, b_i \in A, i = 1, 2, \cdots, n$ 不是唯一的选择, n 是一个自然数.

注记 简单地说, 一个代数 A 上的一阶微积分是一个 A-双模 $\Omega^1(A)$, 连同一个导子 $d: A \longrightarrow \Omega^1(A)$ 使得 $\Omega^1(A) = \mathrm{span}\{xd(a)y \mid x, a, y \in A\}$.

代数 A 上的两个一阶微积分 $(\Gamma, d), (\Gamma', d')$ 称为是相同的, 如果存在一个双模同构 $f: \Gamma \longrightarrow \Gamma'$ 使得 $f(d(a)) = d'(a)$ 对任意 $a \in A$.

设 A 是一个代数带有乘法 $m: A \otimes A \longrightarrow A$. 令 $A^2 = \{q \in A \otimes A \mid mq = 0\} = \mathrm{Ker}(m)$. 显然 A^2 是 $A \otimes A$ 的线性子空间. 那么 A^2 是 A 上的双模, 模结构为
$$c\left(\sum_i^n a_i \otimes b_i\right) = \sum_i^n ca_i \otimes b_i, \quad \left(\sum_i^n a_i \otimes b_i\right)c = \sum_i^n a_i \otimes b_i c$$

对任意 $b \in A$, 设 $D(b) = 1 \otimes b - b \otimes 1$. 易得, $D: A \longrightarrow A^2$ 是线性映射. 进一步, A^2 中的每个元素可以表示为形式: $\sum_i a_i D(b_i)$. 事实上, 设 $\alpha = \sum_i a_i \otimes b_i \in A^2$, 有
$$\alpha = \sum_i a_i(1 \otimes b_i - b_i \otimes 1) + \sum_i a_i b_i \otimes 1 = \sum_i a_i D(b_i).$$

因此, (A^2, D) 是 A 上的一阶微积分.

命题 6.1.18 设 N 是 A^2 的一个子模, $\Gamma = A^2/N$, $\pi: A^2 \longrightarrow \Gamma$ 是经典投射, $d = \pi \circ D$. 那么 (Γ, d) 是 A 上的一阶微积分. 任何 A 上的一阶微积分都具有这种形式.

6.1.5 罗–巴斯特方程

定义 6.1.19 罗–巴斯特(Rota-Baxter) 代数 设 $\lambda \in k$, 一个具有权 λ 的代数 A 上的罗–巴斯特算子是一个线性映射 $P : A \longrightarrow A$ 满足下面等式：

$$P(x)P(y) = P(xP(y)) + P(P(x)y) + \lambda P(xy)$$

对任意 $x, y \in A$.

例 6.1.20 (i) **积分** 设 $A = \mathrm{Cont}(\mathbb{R})$ 是 $\mathbb{R}$ 上的连续函数代数. 定义 $P : A \longrightarrow A$ 为 $P[f](x) = \int_0^x f(t)dt$. 那么 P 是权 0 的罗–巴斯特算子.

事实上, 设 $F(x) := P[f](x) = \int_0^x f(t)dt$, $G(x) := P[g](x) = \int_0^x g(t)dt$, 那么由积分分部法则, 有

$$\int_0^x F(t)G'(t)dt = F(x)G(x) - \int_0^x F'(t)G(t)dt \quad (F(0) = G(0) = 0).$$

即

$$P[P[f]g](x) = P[f](x)P[g](x) - P[fP[g]](x).$$

(ii) **和式** 在一类合适的函数上, 定义

$$P[f](x) := \sum_{n \geq 1} f(x+n).$$

那么, P 是权 1 的罗–巴斯特算子.

事实上, 可以简单计算如下：

$$\begin{aligned}
P[f](x)P[g](x) &= \left(\sum_{n \geq 1} f(x+n)\right)\left(\sum_{m \geq 1} g(x+m)\right) \\
&= \sum_{n \geq 1, m \geq 1} f(x+n)g(x+m) \\
&= \left(\sum_{m > n \geq 1} + \sum_{n > m \geq 1} + \sum_{m=n \geq 1}\right) f(x+n)g(x+m) \\
&= \sum_{m \geq 1}\left(\sum_{k \geq 1, k+m=n} f(x+k+m)\right) g(x+m) \\
&\quad + \sum_{n \geq 1}\left(\sum_{k \geq 1, k+n=m} g(x+k+n)\right) f(x+n) + \sum_{n \geq 1} f(x+n)g(x+n) \\
&= P(P(f)g)(x) + P(fP(g))(x) + P(fg)(x).
\end{aligned}$$

(iii) **部分和** 设 A 是取值在 k 上的序列集合, 那么 A 是代数具有分项和、积与数乘. 定义
$$P: A \longrightarrow A, \quad (a_1, a_2, \cdots) = (a_1, a_1 + a_2, \cdots),$$
那么, P 是权 1 的罗-巴斯特算子.

(iv) **矩阵** 设 A 是上三角 n 阶矩阵代数 $M_{n\times n}(k)$. 定义
$$P: A \longrightarrow A, \quad P(c_{kl})_{ij} = \delta_{ij} \sum_{k \geqslant i} c_{ik},$$
那么, P 是权 -1 的罗-巴斯特算子.

(v) **数量积** 设 A 是代数. 对任意 $\lambda \in k$, 定义
$$P_\lambda : A \longrightarrow A, \quad x \mapsto -\lambda x \quad \text{对任意 } x \in A,$$
那么, (A, P_λ) 是权 λ 的罗-巴斯特代数. 尤其, **id** 是权 -1 的罗-巴斯特算子. 任意代数是罗-巴斯特代数.

(vi) 设 $A = \mathbb{C}[t^{-1}, t]$ 是 Laurent 级数 $\sum_{n=-k}^{\infty} a_n t^n$, $k \geqslant 0$ 构成的代数. 定义
$$P_\lambda : A \longrightarrow A, \quad \sum_{n=-k}^{\infty} a_n t^n \mapsto \sum_{n=-k}^{-1} a_n t^n,$$
那么, (A, P_λ) 是权 -1 的 Rota-Baxter 代数.

(vii) **经典杨-巴斯特方程 (CYB)** 设 $\mathfrak{g}$ 是李代数带有自对偶 $\mathfrak{g}^* := \text{Hom}(\mathfrak{g}, k) \cong \mathfrak{g}$. 那么 $\mathfrak{g}^{\otimes 2} \cong \mathfrak{g} \otimes \mathfrak{g}^* \cong \text{End}(\mathfrak{g})$. 设 $r_{12} \in \mathfrak{g}^{\otimes 2}$ 是反对称. 那么 r_{12} 是下面经典杨-巴斯特方程的解 (r-矩阵):
$$\text{CYB}(r) := [r_{12}, r_{13}] + [r_{12}, r_{23}] + [r_{13}, r_{23}] = 0$$
当且仅当对应 $P \in \text{End}(\mathfrak{g})$ 是一个 (李代数) 权 0 的罗-巴斯特算子:
$$[P(x), P(y)] = P[P(x), y] + P[x, P(y)].$$

(viii) **结合杨-巴斯特方程** 设 A 是结合代数, 且 $r := \sum_i u_i \otimes v_i \in A \otimes A$ 是下面结合杨-巴斯特方程的解:
$$r_{13} r_{12} - r_{12} r_{23} + r_{23} r_{13} = 0.$$
那么 $P_r(x) := \sum_i u_i x v_i$ 定义了一个权 0 的罗-巴斯特算子.

6.2 Hopf 代数方法

本节主要针对 6.1 节所列出的非线性方程, 给出对应的 Hopf 代数方法与理论.

作为预备知识, 设 $\{m_1, m_2, \cdots, m_n\}$ 是 M 的一组基且 $\{p_1, p_2, \cdots, p_n\}$ 是对应的 M^* 的一组对偶基使得 $\langle p^i, m_j \rangle = \delta_j^i$, 对任意 $i,j = 1, 2, \cdots, n$, 这里 δ_j^i 是 Kronecker 符号. 那么 $\{e_j^i = p^i \otimes m_j \mid i,j = 1, 2, \cdots, n\}$ 和 $\{c_j^i = m_j \otimes p^i \mid i,j = 1, 2, \cdots, n\}$ 分别是 $\mathrm{End}(M) \cong M^* \otimes M$ 和 $\mathrm{End}(M^*) \cong M \otimes M^*$ 的自由基. 同构分别如下给出:

$$e_j^i(m_k) = \delta_k^i m_j, \quad c_j^i(p_k) = \delta_j^k p_i.$$

$\mathrm{End}(M) \cong M^* \otimes M$ 同构于 n 阶矩阵代数 $M_n(k)$, 且其上的代数结构为

$$e_j^i e_l^k = \delta_l^i e_j^k, \quad \sum_{i=1}^n e_i^i = 1.$$

类似地, $\mathrm{End}(M^*) \cong M \otimes M^*$ 同构于 n 阶余矩阵余代数 $M^n(k)$, 且其上的余代数结构为

$$\Delta(c_j^i) = \sum_u c_u^i \otimes c_j^u, \quad \varepsilon(c_j^i) = \delta_j^i$$

对任意 $j, k = 1, 2, \cdots, n$.

如果取 $M = k^n$, 我们得到矩阵代数 $M_n(k)$ 和余矩阵余代数 $M^n(k)$ 和经典基. e_j^i 就等同于 (j, i)-位置为 1, 而其他位置为 0 的初等矩阵.

一个线性映射 $R: M \otimes M \longrightarrow M \otimes M$ 可以由它的矩阵表示, 这个是 n^2 阶矩阵带有数 $x_{uv}^{ij} \in k$, 这里 $i, j, u, v = 1, 2, \cdots, n$, 即

$$R(m_k \otimes m_l) = x_{kl}^{ij} m_i \otimes m_j$$

对任意 $k, l = 1, 2, \cdots, n$, 或

$$R = \sum_{i,j,k,l} x_{kl}^{ij} e_i^k \otimes e_j^l, \quad \text{或} \quad R = x_{kl}^{ij} e_i^k \otimes e_j^l,$$

我们经常写为 $R = (x_{kl}^{ij})$.

6.2.1 对于 Hopf 方程的 FRT 理论

我们给出 Hopf 方程的矩阵形式如下.

命题 6.2.1 设有限维向量空间 M 的一组基为 $\{v_1, v_2, \cdots, v_n\}$. 设 $R, S \in \mathrm{End}(M \otimes M)$ 对应的矩阵分别为

$$R(v_k \otimes v_l) = x_{kl}^{ij} v_i \otimes v_j, \quad S(v_k \otimes v_l) = y_{kl}^{ij} v_i \otimes v_j$$

6.2 Hopf 代数方法

对任意 $k,l = 1,2,\cdots,n$, 这里 $(x_{kl}^{ij}), (y_{kl}^{ij})$ 是两组数. 那么有

$$R^{23}S^{13}S^{12} = S^{12}R^{23}$$

当且仅当

$$x_{vu}^{ij}y_{\beta k}^{pu}y_{ql}^{\beta v} = x_{lk}^{\alpha j}y_{q\alpha}^{\rho i}$$

对任意 $i,j,k,l,p,q = 1,2,\cdots,n$. 尤其, R 是 Hopf 方程的解当且仅当

$$x_{vu}^{ij}x_{\beta k}^{pu}x_{ql}^{\beta v} = x_{lk}^{\alpha j}x_{q\alpha}^{pi}$$

对任意 $i,j,k,l,p,q = 1,2,\cdots,n$.

接下来, 我们来看如何由 Hopf 方程的解来构造不带余单位元的双代数.

命题 6.2.2 设 A 是代数, $R \in A \otimes A$ 是 Hopf 方程的一个可逆解. 定义

$$\Delta_l : A \longrightarrow A \otimes A, \quad a \mapsto R(1 \otimes a)R^{-1},$$

那么 $(A, 1_A, \Delta_l)$ 是一个不带余单位的双代数.

类似地, 我们有不带余单位的双代数 $(A, 1_A, \Delta_r)$, 这里

$$\Delta_r : A \longrightarrow A \otimes A, \quad a \mapsto R^{-1}(a \otimes 1)R.$$

命题 6.2.3 设 M 是一个向量空间, $m : M \otimes M \longrightarrow M, v \otimes w \mapsto v \cdot w$ 是一个线性映射. 设

$$R = R_m : M \otimes M \longrightarrow M \otimes M, \quad R(v \otimes w) = v \otimes w \cdot v.$$

那么 R_m 是五角方程的解当且仅当 (M,m) 是一个结合代数.

定理 6.2.4 在线性空间范畴上的所有张量范畴结构 (没有单位对象) 和每个线性空间上的五角方程的所有双射解之间存在一一对应.

定理 6.2.5 设 H 是双代数, 且 $(M,\cdot,\rho) \in {}_H\mathcal{H}^H$ 是左右 H-Hopf 模. 那么映射

$$R = R_{(M,\cdot,\rho)} : M \otimes M \longrightarrow M \otimes M, \quad R(m \otimes n) = n_{(1)} \cdot m \otimes n_0$$

对任意 $m,n \in M$, 是 Hopf 方程的解. 进一步, 如果 H 是交换的 Hopf, 那么 R 是交换的解.

注记 6.2.6 如果 $(M,\cdot,\rho) \in {}_H\mathcal{H}^H$ 是左右 H-Hopf 模, 那么映射

$$R' = \tau R_{(M,\cdot,\rho)}\tau : M \otimes M \longrightarrow M \otimes M, \quad R'(m \otimes n) = m_0 \otimes m_{(1)} \cdot n$$

对任意 $m,n \in M$ 是五角方程的解.

引理 6.2.7 设 M 是有限维向量空间, 且 $\{m_1, m_2, \cdots, m_n\}$ 是 M 的一组基. 设 C 是一个余代数. 设 $\{c_l^v \mid v, l = 1, \cdots, n\}$ 是 C 中的一组元素, 考虑线性映射

$$\rho: M \longrightarrow M \otimes C, \quad \rho(m_l) = m_v \otimes c_l^v,$$

那么 (M, ρ) 是右 C-余模当且仅当矩阵 (c_l^v) 是余乘法, 即

$$\Delta(c_k^j) = c_u^j \otimes c_k^u, \quad \delta(c_k^j) = \delta_k^j$$

对任意 $j, k = 1, \cdots, n$.

令 $B = (c_l^v)$, 上面形式可写为

$$\Delta(B) = B \otimes B, \quad \varepsilon(B) = I_n.$$

引理 6.2.8 设 (C, Δ, ε) 是余代数. 那么, 在张量代数 $(T(C), m, \mu)$ 上存在唯一双代数结构 $(T(C), m, \mu, \overline{\Delta}, \overline{\varepsilon})$ 使得 $\overline{\Delta}(c) = \Delta(c), \overline{\varepsilon}(c) = \varepsilon(c)$ 对任意 $c \in C$. 另外, 包含映射 $i: C \hookrightarrow T(C)$ 是一个余代数同态. 进一步, 如果 M 是一个向量空间且 $\psi: C \otimes M \longrightarrow M, (c, m) \mapsto c \cdot m$ 是一个线性映射, 那么在 M 上存在唯一一个左 $T(C)$-模结构 $\overline{\psi}: T(C) \otimes M \longrightarrow M$ 使得 $\overline{\psi}(c, m) = c \cdot m$ 对任意 $c \in C, m \in M$.

引理 6.2.9 设 H 是双代数, 且 $(M, \cdot)$ 是左 H-模, (M, ρ) 是右 H-余模. 那么下面集合

$$\{h \in H \mid \rho(h \cdot m) = \Delta(h)\rho(m)h(1), \forall m \in M\}$$

是 H 的子代数.

引理 6.2.10 设 H 是双代数, 且 $(M, \cdot)$ 是左 H-模, (M, ρ) 是右 H-余模. 如果 $I \triangleleft H$ 是 H 的双理想使得 $I \cdot M = 0$, 那么具有自然结构下 $(M, \cdot')$ 是左 H/I-模, (M, ρ') 是右 H/I-余模, 且 $R_{(M, \cdot', \rho')} = R_{(M, \cdot, \rho)}$.

定理 6.2.11 设 M 是有限维向量空间, 且 $R \in \text{End}(M \otimes M)$ 是 Hopf 方程的解.

(i) 存在一个双代数 $B(R)$ 使得 M 有一个 $B(R)$-Hopf 模 $(M, \cdot, \rho)$ 和 $R = R_{(M, \cdot, \rho)}$.

(ii) 双代数 $B(R)$ 具有如下泛性质: 如果 H 是一个双代数使得 $(M, \cdot', \rho') \in {}_H\mathcal{H}^H$ 和 $R = R_{(M, \cdot, \rho)}$, 那么存在唯一双代数映射 $f: B(R) \longrightarrow H$ 使得 $\rho' = (1 \otimes f)\rho$. 进一步, $a \cdot m = f(a) \cdot' m$ 对任意 $a \in B(R), m \in M$.

(iii) 如果 R 是交换的, 那么存在一个交换双代数 $\overline{B}(R)$ 使得 M 有一个 $\overline{B}(R)$-Hopf 模结构 $(M, \cdot', \rho')$ 和 $R = R_{(M, \cdot', \rho')}$.

这里简单提示: 设 $\{m_1, m_2, \cdots, m_n\}$ 是 M 的一组基且 $R = (x_{uv}^{ji})$ 是矩阵表达式, 即

$$R(m_v \otimes m_u) = x_{vu}^{ij} m_i \otimes m_j$$

6.2 Hopf 代数方法

对任意 $u, v = 1, 2, \cdots, n$.

设 $(C, \Delta, \varepsilon) = M^n(k)$ 是阶为 n 的余矩阵余代数. 设 $\{c_j^i \mid i, j = 1, 2, \cdots, n\}$ 是 C 的经典基. 那么有

$$\Delta(c_j^i) = c_u^i \otimes c_j^u, \quad \varepsilon(c_j^i) = \delta_j^i$$

对任意 $j, k = 1, 2, \cdots, n$.

设

$$\rho: M \longrightarrow M \otimes C, \quad \rho(m_l) = m_v \otimes c_l^v$$

对任意 $l = 1, 2, \cdots, n$. 由引理 6.2.7 可知 (M, ρ) 是右 C-余模. 再由引理 6.2.8, 可知 $T(C)$ 上有一个双代数结构, 包含映射 $i: C \hookrightarrow T(C)$ 是一个余代数同态. M 有右 $T(C)$-余模结构:

$$\overline{\rho}: M \xrightarrow{\rho} M \otimes C \xrightarrow{\mathbf{id} \otimes i} M \otimes T(C).$$

定理 6.2.12 设 H 是双代数, 且 $(M, \cdot, \rho) \in {}_H\mathcal{YD}^H$ 是左右 Yetter-Drinfel'd 模. 那么映射

$$R = R_{(M, \cdot, \rho)}: M \otimes M \longrightarrow M \otimes M, \quad R(m \otimes n) = n_{(1)} \cdot m \otimes n_0$$

对任意 $m, n \in M$, 是量子杨–巴斯特方程的解. 进一步, 如果 H 是一个 Hopf 代数, 那么 R 是双射.

定义 6.2.13 设 (Γ, d) 是 Hopf 代数 $(A, \Delta, \varepsilon, S)$ 上的一阶微积分. 我们说 (Γ, d) 是:

(i) **左余不变** 如果对任意 $a_i, b_i \in A$ 有

$$\left(\sum_i a_i \mathbf{d}(b_i) = 0\right) \Longrightarrow \left(\sum_i \Delta(a_i)(\mathbf{id} \otimes d)\Delta(b_i) = 0\right).$$

(ii) **右余不变** 如果对任意 $a_i, b_i \in A$ 有

$$\left(\sum_i a_i \mathbf{d}(b_i) = 0\right) \Longrightarrow (\sum_i \Delta(a_i)(d \otimes \mathbf{id})\Delta(b_i) = 0).$$

(iii) **双余不变** 如果它既是左余不变又是右余不变的.

定义 6.2.14 设 $(A, \Delta, \varepsilon, S)$ 是 Hopf 代数, Γ 是 A-双模.

(i) 称 Γ 是左余不变 A-双模, 如果 (Γ, Δ_L) 是左 A-余模且满足

$$\Delta_L(ax) = \Delta(a)\Delta_L(x), \quad \Delta_L(xa) = \Delta_L(x)\Delta(a)$$

对任意 $a \in A, x \in \Gamma$.

(ii) 称 Γ 是右余不变 A-双模, 如果 (Γ, Δ_R) 是右 A-余模且满足

$$\Delta_R(ax) = \Delta(a)\Delta_R(x) \quad \text{和} \quad \Delta_R(xa) = \Delta_R(x)\Delta(a)$$

对任意 $a \in A, x \in \Gamma$.

(iii) 称 Γ 是双余不变 A-双模, 如果它既是左余不变 A-双模又是右余不变 A-双模, 而且有

$$(\mathbf{id} \otimes \Delta_R)\Delta_L = (\Delta_L \otimes \mathbf{id})\Delta_R \quad \text{和} \quad (\Delta_L \otimes \mathbf{id})\Delta_R = (\mathbf{id} \otimes \Delta_R)\Delta_L.$$

设 $(\Gamma_1, \Delta_{L_1}, \Delta_{R_1}), (\Gamma_2, \Delta_{L_2}, \Delta_{R_2})$ 是两个双余不变 A-双模. 我们说线性映射 $f : \Gamma_1 \longrightarrow \Gamma_2$ 是双余不变双模映射, 如果成立等式:

$$(\mathbf{id} \otimes f)\Delta_{L_1} = \Delta_{L_2} f \quad \text{和} \quad (f \otimes \mathbf{id})\Delta_{R_1} = \Delta_{R_2} f.$$

一阶微积分 (Γ, d) 的左 (右) 余不变可推出 (Γ, d) 是左 (右) 余不变 A-双模且 d 是双余不变双模映射.

为了建立任意左 (右) 余不变微积分, 设 $T_1 : A \otimes A \longrightarrow A \otimes A, a \otimes b \mapsto (a \otimes 1)\Delta(b)$ 和 $T_2 : A \otimes A \longrightarrow A \otimes A, a \otimes b \mapsto (1 \otimes a)\Delta(b)$. 同时考虑伴随余作用: Ad: $A \longrightarrow A \otimes A, a \mapsto \sum a_2 \otimes S(a_1)a_3$. 我们称 A 的线性子空间 B 是 Ad-不变的, 如果 $\mathrm{Ad}(B) \subseteq B \otimes A$.

定理 6.2.15 设 $(A, \Delta, \varepsilon, S)$ 是 Hopf 代数. 一个 A 上的微积分 (Γ, d) 是一个左 (右) 余不变的当且仅当存在 A 的右理想 $R \subseteq \mathrm{Ker}(\varepsilon)$ 使得 $\Gamma = A^2/N$, $N = T_1^{-1}(A \otimes R)$ $(N = T_2^{-1}(R \otimes A))$, 且 $d = \pi D$, 这里 $\pi : A^2 \longrightarrow A^2/N$ 为投射. 进一步, (Γ, d) 是双余不变的当且仅当 R 是 Ad-不变的.

注记 这个定理是构造 Hopf 代数上的双余不变微积分的有用工具.

如果 (Γ, d) 是 A 上的一阶微积分, 那么外代数 $\Omega(A) = \bigoplus_{n=0} \Omega^n(A)$ (这里 $\Omega^0(A) = A, \Omega^1(A) = \Gamma$). 算子 d 可以唯一扩张到整个 $\Omega(A)$ 上外导子, 满足

$$(\mathbf{id} \otimes d)\overline{\Delta}_L = \overline{\Delta}_L d \quad \text{和} \quad (d \otimes \mathbf{id})\overline{\Delta}_R = \overline{\Delta}_R d,$$

这里 $\overline{\Delta}_{L,R}$ 是 $\Delta_{L,R}$ 到 $\Omega(A)$ 上的扩张. Γ 上的外代数定义如下: 我们说元素 $\omega \in \Gamma$ 是左不变的, 如果 $\Delta_L(\omega) = 1 \otimes \omega$. 类似地, 有右不变的概念. 一次形 $\omega \in \Omega^1(A) = \Gamma$ 是双不变的, 如果它既是左不变的又是右不变的. 对任意 $p \in \Gamma$, 有唯一的表达式: $p = \sum_i \omega_i a^i$, 这里 $a^i \in A$, $\{\omega_i\}$ 是所有左 (右) 不变形空间的一组基.

类似地, 对任意 $x \in \Gamma \otimes_A \Gamma$, 可以唯一写成 $x = \sum \omega_i \otimes_A \eta_j a^{ij}$, 这里 $a^{ij} \in A$, $\{\omega_i\}(\{\eta_j\})$ 是所有左 (右) 不变形空间的一组基. 因此, 我们可以定义形式外积 (wedge product): $\Gamma \wedge \Gamma := \Gamma \otimes_A \Gamma/\mathrm{Ker}(\mathbf{id} - t)$, 这里 $t : \Gamma \otimes_A \Gamma \longrightarrow \Gamma \otimes_A \Gamma, \omega \otimes \eta \mapsto \eta \otimes \omega$

6.2 Hopf 代数方法

对所有左 (右) 不变 $\omega(\eta)$. 这个外积定义可以扩张到整个 $\Omega^n(A)$ 上. d 也可以通过双不变一次型 θ 扩张, 详细地, 有

$$d(\rho) = \theta \wedge \rho - (-1)^{\partial \rho} \rho \wedge \theta,$$

这里 $\rho \in \Omega(A)$, $\partial \rho$ 表示 ρ 的次数.

定义 6.2.16 设 A 是一个 Hopf 代数, $\Omega(A) = \bigoplus_{n=0}^{\Omega^n}$ 是 $\mathbb{Z}_2$-分次双代数 (Hopf 代数) 使得 $\Omega^0 = A$. 如果存在一个次数 1 的双代数 (Hopf 代数) 同态 $d: \Omega \longrightarrow \Omega$, 服从分次莱布尼茨 (Leibniz) 法则, 且使得 $d^2 = 0$. 如果 Ω^n 中元素都具有形式: $a_0 d(a_1) \cdots d(a_n)$, 这里 $a_0, a_1, \cdots, a_n \in A$. 那么我们说 (Ω, d) 是 A 的一个外微分双代数 (Hopf 代数). 微分 d 在张量积 $\Omega \otimes \Omega$ 上的作用定义为

$$d(\omega \otimes \eta) = d(\omega) \otimes \eta + (-1)^{\partial \omega} \omega \otimes d(\eta)$$

对任意 $\omega \in \Omega$ 和齐次元 $\eta \in \Omega^n$ 具有次数 $\partial \omega$.

定理 6.2.17 设 (Γ, d) 是 Hopf 代数 A 上的一个双余不变一阶微积分. 那么 $(\Omega(A), d)$ 是 A 的一个外 Hopf 代数. 反之, 如果 $(\Omega(A), d)$ 是 Hopf 代数 A 的一个外双代数, 那么 Ω^1 是 A 上的一个双余不变双模使得 d 为余模映射.

推论 6.2.18 设 (Γ, d) 是 Hopf 代数 A 上的双不变微积分. 那么 $\Omega(A)$ 的上同调 $(H^*(\Omega(A)), d)$ 是一个 Hopf 代数.

定义 6.2.19 设 L 是具有双射反对极的 Hopf 代数. 那么 L 的 **Heisenberg 偶**是一个冲积 $\mathcal{H}(L) = L \# L^*$, 具有乘法

$$(h \# h^*)(g \# g^*) = \sum h_2 g \# h^* (h_1 \cdot g^*) = \sum h_2 g \# h^* \langle g^*, h_1 \rangle$$

对任意 $h, g \in L, h^*, g^* \in L^*$.

作为向量空间, 我们有同构: $\phi : \mathcal{H}(L) = L \# L^* \longrightarrow \mathrm{End}(L) = \mathrm{Mat}_{\dim(L)}(k)$, $\phi(1 \# 1^*)(g) = \langle 1^*, g \rangle 1$. 但这不是代数同构. 然而, 我们有如下命题.

命题 6.2.20 设 L 是具有双射反对极的有限维 Hopf 代数. 那么存在一个代数同构: $\mathcal{H}(L) \cong \mathrm{Mat}_{\dim(L)}(k)$.

证明 因为 L 是有限维 Hopf 代数, 有范畴等价

$$T : {}^L\mathrm{Mod}_L \longrightarrow \mathrm{Mod}_{\mathcal{H}(L)}, \quad T(M) = M,$$

这里 $\mathcal{H}(L)$ 在 M 上的作用为 $m \cdot (l \# l^*) = \sum \langle l^*, m_{(-1)} \rangle m_0 \cdot l$ 对任意 $l \in L, l^* \in L^*, m \in M$.

由于 L 有双射反对极, 我们有下面范畴等价:

$$\mathrm{Mod}_{\mathcal{H}(L)} \cong {}^L\mathrm{Mod}_L \cong \mathrm{Mod}_k,$$

即 $\mathcal{H}(L)$ Morita 等价于 k. 那么由 Morita 理论可知, 存在一个代数同构: $\mathcal{H}(L)\mathrm{Mat}_n(k)$. 考虑到 $\dim(\mathcal{H}(L)) = \dim(L)^2$, 有 $n = \dim(L)$.

6.3 进一步研究问题

6.2 节的思想方法可以用于第 4 章中介绍的各类 Hopf 型代数中, 因此有大量问题可进行研究. 例如: 偏作用方面 [32, 67]、双李代数胚 [123]、千年问题中的 PNP 问题 [137]、一些特殊群产生的量子群 [151, 152]、Tannaka-Stinespring-Tatsuuma 对偶理论 [179]、Hopf 代数的 K-理论 [187]、辫子李代数方面 [211]、环论方面 Kegel 定理 [228] 和超群胚方面 [237] 等.

第 7 章 范畴中的基本概念

本章主要介绍与 Hopf 型代数相关的范畴论中基本的概念与例子 (参考 [122, 185]).

7.1 范畴与函子

本节主要介绍范畴、函子、自然变换、伴随函子和范畴等价与同构等概念.

定义 7.1.1 一个范畴 $\mathcal{C}$ 由下面参数组成:

(1) 一个对象类 $|\mathcal{C}| = \mathcal{C}_0 = \mathcal{C}$, 记为: $X, Y, Z, \cdots$;

(2) 对任意两个对象 X, Y, 一个态射集合 $\mathrm{Hom}_\mathcal{C}(X,Y) = \mathrm{Hom}(X,Y) = \mathcal{C}(X,Y)$;

(3) 对任意三个对象 X, Y, Z, 态射合成律为

$$\circ : \mathrm{Hom}(X,Y) \times \mathrm{Hom}(Y,Z) \longrightarrow \mathrm{Hom}(X,Z), \quad (f,g) \mapsto g \circ f;$$

(4) 对任意对象 X, 其上的单位态射, 记为 1_X 或 X;

这些数据满足下面条件:

(5) 对任意对象 X, Y, Z, U 和态射 $f \in \mathrm{Hom}(X,Y), g \in \mathrm{Hom}(Y,Z), h \in \mathrm{Hom}(Z,U)$, 有

$$h \circ (g \circ f) = (h \circ g) \circ f;$$

(6) 对任意对象 X, Y, Z 和态射 $f \in \mathrm{Hom}(X,Y), g \in \mathrm{Hom}(Y,Z)$, 有

$$Y \circ f = f, \quad g \circ Y = g.$$

例 7.1.2 (1) 范畴 **Set**. 其对象是集合, 态射是集合之间的映射.

(2) 设 k 是交换环, 范畴 $\mathcal{M}_k$. 其对象是 k-模, 态射是 k-线性映射.

(3) 设 A 是非交换环, 范畴 $_A\mathcal{M}$. 其对象是左 A-模, 态射是 A-模映射. 类似地, 有范畴 $\mathcal{M}_A$, 范畴 $_A\mathcal{M}_B$, 这里 B 是另一个环.

(4) 范畴 **Top**, 其对象是拓扑空间, 态射是连续映射. 范畴 $\mathbf{Top}_0$, 其对象是有点拓扑空间, 即带有固定基点的拓扑空间, 态射是保持基点的连续映射.

(5) 范畴 **Grp**, 其对象是群, 态射是群同态. 范畴 **Grpoi**, 其对象是群胚, 态射是群胚同态. 范畴 **Ab**, 其对象是交换群, 态射是群同态. 尤其有: $\mathbf{Ab} = \mathcal{M}_\mathbb{Z}$.

(6) 范畴 **Rng**, 其对象是环, 态射是环同态.

(7) 范畴 $\mathbf{Alg}_k$, 其对象是 k-代数, 态射是代数同态. 尤其有: $\mathbf{Alg}_k = \mathbf{Rng}$.

(8) 上面的例子都是具体范畴：对象是集合带有附加结构，它们到范畴 **Set** 有忠实的遗忘函子. 一个非具体范畴如下：设 M 是半群，可以作为一个范畴带有一个对象 $*$，这里 $\mathrm{Hom}(*,*) = M$.

(9) 平凡范畴只有一个对象 $*$，一个态射，即 $*$ 的恒等态射.

(10) 另一个非具体范畴可以考虑为：对象类是 $\mathbb{N}_0$，态射 $\mathrm{Hom}(n,m) = M_{n,m}(k)$，即所有 $n \times m$ 矩阵，这里 k 是交换环.

(11) 如果 $\mathcal{C}$ 是范畴，有反范畴 $\mathcal{C}^{\mathrm{op}}$，其对象不变，态射为 $\mathrm{Hom}_{\mathcal{C}^{\mathrm{op}}}(X,Y) = \mathrm{Hom}_{\mathcal{C}}(Y,X)$.

(12) 如果 $\mathcal{C}, \mathcal{D}$ 是范畴，有积范畴 $\mathcal{C} \times \mathcal{D}$.

(13) **S-分次范畴** 设 S 是集合. 一个 S-分次 k-模是一个 k-模连通一个子模直和分解
$$M = \bigoplus_{s \in S} M_s.$$
这意味着 $m \in M$ 有唯一分解 $m = \sum_{s \in S} m_s$，m_s 称为 m 的次数为 s 的齐次分量. 记 $\deg(m_s) = s$. 两个 S-分次模 M, N 之间的映射 f 称为是分次的，如果它保持次数，即 $f(M_s) \subset N_s$. 我们记 S-分次范畴为 $\mathbf{gr}^S$.

(14) **余模范畴** 设 C 是余代数，记右 C-余模范畴为：$\mathcal{M}^C$. 类似地，有 $^C\mathcal{M}$ 和 $^C\mathcal{M}^C$.

关于余代数，设 k 为交换环，例如：

(i) 设 S 是一个非空集合，$C = kS$ 是基为 S 的自由 k-模. 那么 $(kS, \Delta, \varepsilon)$ 是一个余代数：
$$\Delta(s) = s \otimes s, \quad \varepsilon(s) = 1.$$

(ii) **可除幂余代数** 设 $C = kS$ 是基为 $\{c_m \mid m \in \mathbb{N}\}$ 的自由 k-模. 那么 (C, Δ, ε) 是一个余代数：
$$\Delta(c_m) = \sum_{i=o}^{m} c_i \otimes c_{m-i}, \quad \varepsilon(c_m) = \delta_{m,0}.$$

(iii) **矩阵余代数** 设 $M^n(k)$ 是基为 $\{e_{ij} \mid i,j = 1,2,\cdots,n\}$ 的 n^2 维自由 k-模. 那么 $(M^n(k), \Delta, \varepsilon)$ 是一个余代数：
$$\Delta(e_{ij}) = \sum_{k=o}^{n} e_{ik} \otimes e_{kj}, \quad \varepsilon(e_{ij}) = \delta_{i,j}.$$

(iv) 设 C 是基为 $\{g_m, d_m \mid m \in \mathbb{N}^*\}$ 的 n^2 维自由 k-模. 那么 (C, Δ, ε) 是一个余代数：
$$\Delta(g_m) = g_m \otimes g_m, \quad \varepsilon(g_m) = 1,$$
$$\Delta(d_m) = g_m \otimes d_m + d_m \otimes g_{m+1}, \quad \varepsilon(d_m) = 0.$$

(v) **Trigonometric 余代数** 设 C 是基为 $\{s,c\}$ 的自由 k-模. 那么 (C,Δ,ε) 是一个余代数:

$$\Delta(s) = s \otimes c + c \otimes s, \quad \varepsilon(s) = 0,$$
$$\Delta(c) = c \otimes c + s \otimes s, \quad \varepsilon(c) = 1$$

定义 7.1.3 设 $\mathcal{C}, \mathcal{D}$ 是范畴. 一个共变函子 $F: \mathcal{C} \longrightarrow \mathcal{D}$ 由下面数据组成:
(1) 对任意对象 $X \in \mathcal{C}$, 我们有一个对象 $FX = F(X) \in \mathcal{D}$;
(2) 对任意 $f \in \text{Hom}_{\mathcal{C}}(X,Y)$, 存在一个态射 $Ff = F(f) \in \text{Hom}_{\mathcal{D}}(FX, FY)$ 满足下面条件:
(3) 对任意 $f \in \text{Hom}_{\mathcal{C}}(X,Y), g \in \text{Hom}_{\mathcal{C}}(Y,Z)$, 有

$$F(g \circ f) = F(g) \circ F(f);$$

(4) 对任意对象 $X \in \mathcal{C}$, 我们有 $F(1_X) = 1_{FX}$.

一个反变函子 $F: \mathcal{C} \longrightarrow \mathcal{D}$ 是一个共变函子 $F: \mathcal{C}^{\text{op}} \longrightarrow \mathcal{D}$. 进一步, 我们有: 函子 $F: \mathcal{C} \longrightarrow \mathcal{D}$ 是反变函子当且仅当 $F^{\text{op}}: \mathcal{C}^{\text{op}} \longrightarrow \mathcal{D}$ 和 $F^{\text{cop}}: \mathcal{C} \longrightarrow \mathcal{D}^{\text{op}}$ 是共变的. 显然, 函子 $F^{\text{op,cop}}: \mathcal{C}^{\text{op}} \longrightarrow \mathcal{D}^{\text{op}}$ 也是一个反变函子.

例 7.1.4 (1) 所有具体范畴到集合范畴有一个遗忘函子.
(2) 设 $\mathcal{C}$ 是范畴, 恒等函子 $1_{\mathcal{C}}: \mathcal{C} \longrightarrow \mathcal{C}, 1_{\mathcal{C}}(X) = X, 1_{\mathcal{C}}(f) = f$ 对任意态射 $f \in \text{Hom}_{\mathcal{C}}(X,Y), \forall X, Y \in \mathcal{C}$.
(3) 设 $\mathcal{C}, \mathcal{D}$ 是范畴, $X \in \mathcal{D}$, 一个常函子 $\mathcal{C} \longrightarrow \mathcal{D}, C \mapsto X, f \in \text{Hom}_{\mathcal{C}}(C,C') \mapsto \text{id}_X$. 事实上, 一个常函子等价于选择了一个对象 $D \in \mathcal{D}$.
(4) 张量函子 $- \otimes -: \mathcal{M}_k \times \mathcal{M}_k \longrightarrow \mathcal{M}_k, (X,Y) \mapsto X \otimes Y$.
(5) 有共变函子 $\pi_1: \textbf{Top}_0 \longrightarrow \textbf{Grp}, (X, X_0) \mapsto \pi_1(X, X_0)$, 这里 $\pi_1(X, X_0)$ 表示 (X, X_0) 的基本群 (fundamental group).
(6) 反变函子 $(-)^*: \mathcal{M}_k \longrightarrow \mathcal{M}_k, X \mapsto X^* = \text{Hom}(X, k)$.

定义 7.1.5 设 $F, G: \mathcal{C} \longrightarrow \mathcal{D}$ 是两个函子. 一个自然变换 $\alpha: F \longrightarrow G$ (有时记为 $\alpha: F \Rightarrow G$), 对每个对象 $C \in \mathcal{C}$, 在 $\mathcal{D}$ 中都有一个态射 $\alpha_C: FC \longrightarrow GC$, 使得对任意 $f \in \text{Hom}_{\mathcal{C}}(X,Y)$ 有下面图交换:

$$\begin{array}{ccc} FX & \xrightarrow{Ff} & FY \\ \alpha_X \downarrow & & \downarrow \alpha_Y \\ GX & \xrightarrow{Gf} & GY \end{array}$$

如果对任意 $X \in \mathcal{C}, \alpha_X$ 是同构的, 我们说 $\alpha: F \longrightarrow G$ 是一个自然同构.

例 7.1.6 (1) 设 $F: \mathcal{C} \longrightarrow \mathcal{D}$ 是个函子, 那么 $1_F: F \longrightarrow F, (1_F)X = 1_{FX}: FX \longrightarrow FX$ 是 F 上的恒等自然变换.

(2) 对一个自然变换 $\alpha: F \longrightarrow G$, 我们也说态射 $\alpha_C: FC \longrightarrow GC$ 在 $\mathcal{C}$ 中自然. 例如: 态射 $X \otimes Y^* \longrightarrow \mathrm{Hom}(Y, X), x \otimes f \mapsto (y \mapsto xf(y))$ 在 X 和 Y 处都自然. 这里 X, Y 都是 k-模.

(3) 经典内射 $\iota_X: X \longrightarrow X^{**}, \iota_X(x)(f) = f(x)$ 对任意 $x \in X, f \in X^*$, 定义了一个自然变换 $\iota: 1_{\mathcal{M}_k} \longrightarrow (-)^{**}$. 如果我们限制到有限生成投射 k-模, 那么 ι 是一个自然同构.

定义 7.1.7 设 $\mathcal{C}, \mathcal{D}$ 是范畴. 如果 $L: \mathcal{C} \longrightarrow \mathcal{D}$ 和 $R: \mathcal{D} \longrightarrow \mathcal{C}$ 是两个函子. 下面说法等价:

(1) L 是 R 的左伴随函子;

(2) R 是 L 的右伴随函子;

(3) (L, R) 是伴随对;

(4) 存在一个自然同构 $\theta_{C,D}: \mathrm{Hom}_{\mathcal{D}}(LC, D) \longrightarrow \mathrm{Hom}_{\mathcal{C}}(C, RD)$ 对任意 $C \in \mathcal{C}, D \in \mathcal{D}$;

(5) 存在两个自然变换 $\eta: 1_{\mathcal{C}} \longrightarrow RL$ (称为单位) 和 $\varepsilon: LR \longrightarrow 1_{\mathcal{D}}$, 使得下面图可换: 对任意 $C \in \mathcal{C}, D \in \mathcal{D}$

$$\begin{array}{ccc} LC \xrightarrow{L\eta_C} LRLC & \quad & RD \xrightarrow{\eta_{RD}} RLRD \\ \parallel \quad \downarrow \varepsilon_{LC} & \quad & \parallel \quad \downarrow R\varepsilon_D \\ LC \xrightarrow{=} LC & \quad & RD \xrightarrow{=} RD \end{array}$$

这意味着: $1_L = \varepsilon L \circ L\eta$ 和 $1_R = R\varepsilon \circ \eta R$.

例 7.1.8 (1) 设 $U: \mathbf{Grp} \longrightarrow \mathbf{Set}$ 是遗忘函子, 它有左伴随函子 $F: \mathbf{Set} \longrightarrow \mathbf{Grp}$, 把每个集合映射到该集合元素生成的自由群.

(2) 设 X 是 k-模. 函子 $- \otimes X$ 是函子 $\mathrm{Hom}(X, -)$ 的左伴随函子.

(3) 设 X 是 (A, B)-双模. 函子 $- \otimes_A X: \mathcal{M}_A \longrightarrow \mathcal{M}_B$ 是函子 $\mathrm{Hom}_B(X, -): \mathcal{M}_B \longrightarrow \mathcal{M}_A$ 的左伴随函子.

(4) 设 $f: B \longrightarrow A$ 是 k-代数同态, 那么标量函子的限制 $R: \mathcal{M}_A \longrightarrow \mathcal{M}_B$ 是诱导函子 $- \otimes_B A: \mathcal{M}_B \longrightarrow \mathcal{M}_A$ 的右伴随函子. 回顾, 对任意 $X \in \mathcal{M}_A, R(X) = X$ 作为 k-模, 而 B 在 $R(X)$ 上的右作用定义为: $x \cdot b = xf(b)$ 对任意 $x \in X, b \in B$.

(5) 设 $k-: \mathbf{Grp} \longrightarrow \mathbf{Alg}_k$ 把任意群 G 映射到群代数 kG. 设 $U: \mathbf{Alg}_k \longrightarrow \mathbf{Grp}$ 把任意 k-代数 A 映射到它的单位群 $U(A)$. 那么 $k-$ 是 U 的左伴随函子.

定义 7.1.9 设 $F: \mathcal{C} \longrightarrow \mathcal{D}$ 是函子. 那么 F 诱导下面态射:

$$F_{C,C'}: \mathrm{Hom}_{\mathcal{C}}(C, C') \longrightarrow \mathrm{Hom}_{\mathcal{D}}(FC, FC'),$$

其在 $C \in \mathcal{C}, D \in \mathcal{D}$ 处自然. 那么 F 称为是

(1) 忠实函子, 如果 $F_{C,C'}$ 是单射;

(2) 全函子, 如果 $F_{C,C'}$ 是满射;

(3) 忠实的全函子, 如果 $F_{C,C'}$ 是双射.

命题 7.1.10 如果 (L,R) 是函子伴随对带有单位 η 和余单位 ε, 那么

(i) L 是忠实的全函子当且仅当 η 是自然同构;

(ii) R 是忠实的全函子当且仅当 ε 是自然同构.

命题 7.1.11 设 S 是集合, k 是交换环. 那么分次模范畴 $\mathbf{gr}^S$ 同构于范畴 $\mathcal{M}^{kS}$.

证明 定义两个互逆函子:

$$F: \mathbf{gr}^S \longrightarrow \mathcal{M}^{kS}, \quad M \mapsto M, \quad 对 \quad m = \sum_{s \in S} m_s, \quad \rho(m) = \sum_{s \in S} m_s \otimes s;$$

$$G: \mathcal{M}^{kS} \longrightarrow \mathbf{gr}^S, \quad M \mapsto M = \sum_{s \in S} M_s, \quad 这里 \quad M_s = \{m \mid \rho(m) = m \otimes s\}.$$

7.2 辫子张量范畴

本节主要介绍张量范畴、对称和辫子张量范畴等概念.

定义 7.2.1 一个**张量范畴**(monoidal 或 tensor category) 是六元素组 $\mathcal{C} = (\mathcal{C}, \otimes, k, a, l, r)$, 这里:

(1) $\mathcal{C}$ 是一个范畴;

(2) k 是 $\mathcal{C}$ 中的对象, 称为**单位对象**;

(3) $-\otimes - : \mathcal{C} \times \mathcal{C} \longrightarrow \mathcal{C}$ 是一个函子, 称为**张量积**;

(4) $a: \otimes \circ (\otimes \times \mathbf{id}) \longrightarrow \otimes \circ (\mathbf{id} \times \otimes)$ 是自然同构, 称为**结合子**;

(5) $l: \otimes \circ (k \times \mathbf{id}) \longrightarrow \mathbf{id}$ 和 $r: \otimes \circ (\mathbf{id} \times k) \longrightarrow \mathbf{id}$ 是自然同构, 分别称为**单位子**.

这意味着, 有下面的同构:

$$a_{M,N,P}: (M \otimes N) \otimes P \longrightarrow M \otimes (N \otimes P);$$
$$l_M: k \otimes M \longrightarrow M; \quad r_M: M \otimes k \longrightarrow M$$

对任意 $M, N, P \in \mathcal{C}$. 我们也要求下面图形交换:

$$\begin{array}{ccccc}
((M \otimes N) \otimes P) \otimes Q & \xrightarrow{a_{M \otimes N, P, Q}} & (M \otimes N) \otimes (P \otimes Q) & \xrightarrow{a_{M,N,P \otimes Q}} & M \otimes (N \otimes (P \otimes Q)) \\
{\scriptstyle a_{M,N,P} \otimes Q} \downarrow & & & & \uparrow {\scriptstyle M \otimes a_{N,P,Q}} \\
(M \otimes (N \otimes P)) \otimes Q & & \xrightarrow{a_{M, N \otimes P, Q}} & & M \otimes ((N \otimes P) \otimes Q)
\end{array}$$

$$(M\otimes k)\otimes N \xrightarrow{a_{M,k,N}} M\otimes (k\otimes N)$$
$$\downarrow r_M\otimes N \qquad\qquad\qquad \downarrow M\otimes l_N$$
$$M\otimes N \xrightarrow{\quad = \quad} M\otimes N$$

例 7.2.2 (1) $(\mathbf{Set}, \times, \{*\})$ 是张量范畴. 这里 $\{*\}$ 是固定的单点集合, 例如: 结合子 a 和单位子 l, r 如下:

$$a_{M,N,P} : (M\times N)\times P \longrightarrow M\times (N\times P); \quad a_{M,N,P}((m,n),p) = (m,(n,p));$$
$$l_M : \{*\}\times M \longrightarrow M; \quad l_M(*,m) = m, \quad r_M : M\times \{*\} \longrightarrow M; \quad r_M(m,*) = m.$$

(2) 对于交换环 k, $({}_k\mathcal{M}, \otimes, k)$ 是张量范畴. 结合子 a 和单位子 l, r 如下: 我们用 $\otimes$ 代替 $\otimes_k$

$$a_{M,N,P} : (M\otimes N)\otimes P \longrightarrow M\otimes (N\otimes P); \quad a_{M,N,P}((m\otimes n)\otimes p) = m\otimes (n\otimes p);$$
$$l_M : k\otimes M \longrightarrow M; l_M(k\otimes m) = km, \quad r_M : M\otimes k \longrightarrow M; \quad r_M(m\otimes k) = mk.$$

但是, 如果我们考虑如下映射:

$$a'_{M,N,P} : (M\otimes N)\otimes P \longrightarrow M\otimes (N\otimes P); \quad a'_{M,N,P}((m\otimes n)\otimes p) = -m\otimes (n\otimes p),$$

这定义了一个自然同构, 但当 k 的特征不等于 2 时, 五角图确实不成立.

(3) 对于交换环 k, $({}_k\mathcal{M}, \times, 0)$ 是张量范畴, 这里 0 是模. 结合子 a 和单位子 l, r 如下

$$a_{M,N,P} : (M\times N)\times P \longrightarrow M\times (N\times P); \quad a_{M,N,P}((m,n),p) = (m,(n,p));$$
$$l_M : 0\times M \longrightarrow M; \quad l_M(0,m) = m, \quad r_M : M\times 0 \longrightarrow M; \quad r_M(m,0) = m.$$

(4) 设 G 是幺半群, $({}_{kG}\mathcal{M}, \otimes, k)$ 是张量范畴.

定义 7.2.3 设 $\mathcal{C}_1, \mathcal{C}_2$ 是两个张量范畴. 从 $\mathcal{C}_1$ 到 $\mathcal{C}_2$ 的张量函子是三元组 (F, φ_0, φ), 这里:

- F 是一个函子;
- $\varphi_0 : k_2 \longrightarrow F(k_1)$ 是 $\mathcal{C}_2$-态射;
- $\varphi : \otimes \circ (F, F) = F\circ \otimes$ 是 $\mathcal{C}_1\times\mathcal{C}_1 \longrightarrow \mathcal{C}_2$ 函子之间的自然变换. 所以, 我们有一族态射

$$\varphi_{M,N} : F(M)\otimes F(N) \longrightarrow F(M\otimes N)$$

使得下图交换: 对任意 $M, N, P \in \mathcal{C}_1$,

$$\begin{array}{ccc}
(F(M)\otimes F(N))\otimes F(P) & \xrightarrow{a_{F(M),F(N),F(P)}} & F(M)\otimes(F(N)\otimes F(Q)) \\
\varphi_{M,N}\otimes F(P)\downarrow & & \downarrow F(M)\otimes\varphi_{P,Q} \\
F(M\otimes N)\otimes F(P) & & F(M)\otimes F(N\otimes Q) \\
\varphi_{M\otimes N,P}\downarrow & & \downarrow \varphi_{M,N\otimes Q} \\
F((M\otimes N)\otimes P) & \xrightarrow{F(a_{M,N,P})} & F(M\otimes(N\otimes P))
\end{array}$$

$$\begin{array}{ccc}
k_2\otimes F(M) & \xrightarrow{l_{F(M)}} & F(M) \\
\varphi_0\otimes F(M)\downarrow & & \downarrow F(l_M) \\
F(k_1)\otimes F(M) & \xrightarrow{\varphi_{k_1,M}} & F(k_1\otimes M)
\end{array}$$

$$\begin{array}{ccc}
F(M)\otimes k_2 & \xrightarrow{r_{F(M)}} & F(M) \\
F(M)\otimes\varphi_0\downarrow & & \downarrow F(r_M) \\
F(M)\otimes F(k_1) & \xrightarrow{\varphi_{M,k_1}} & F(M\otimes k_1)
\end{array}$$

如果 φ_0 是同构且 φ 是自然同构, 那么我们说 F 是强张量函子. 如果 φ_0 和 φ 是恒等映射, 那么我们说 F 是严格张量函子.

例 7.2.4 (1) 考虑函子 $k\text{-}: \mathbf{Sets} \longrightarrow {}_k\mathcal{C}$, 将集合 X 映射到向量空间 kX, 带有基 X. 对于函子 $f: X \longrightarrow Y$, 有线性映射

$$\overline{f}\left(\sum_{x\in X}a_x x\right) = \sum_{x\in X}a_x f(x).$$

同构 $\varphi_0: k \longrightarrow k*, a \mapsto a*$.

$$\varphi_{X,Y}: kX\otimes kY \longrightarrow k(X\times Y), \quad x\otimes y \mapsto (x,y)$$

对任意 $x\in X, y\in Y$. 因此, 线性函子 k-是强张量函子.

(2) 设 G 是幺半群. 遗忘函子 $U: {}_{kG}\mathcal{M} \longrightarrow {}_k\mathcal{M}$ 是强张量函子. 映射 $\varphi_0: k \longrightarrow U(k) = k$, $\varphi_{M,N}: U(M)\otimes U(N) = M\otimes N \longrightarrow U(M\otimes N) = M\otimes N$ 是恒等映射. 所以, U 是严格张量函子.

(3) $\text{Hom}(-, k): \mathbf{Set}^{\text{op}} \longrightarrow \mathcal{M}_k$ 是张量函子.

定义 7.2.5 一个张量范畴 $(\mathcal{C}, \otimes, I)$ 是**辫子张量范畴**, 如果存在一个自然同构 $\sigma: \otimes \Rightarrow \otimes\tau$, 这里 $\tau: \mathcal{C}\times\mathcal{C} \longrightarrow \mathcal{C}\times\mathcal{C}, (M,N) \mapsto (N,M)$ 是交换函子. 自然意味着,

对任意 $M, N \in \mathcal{C}$, 存在一个同构 $\sigma_{M,N}: M \otimes N \longrightarrow N \otimes M$ 使得下图交换: 对任意 $f: M \longrightarrow M', g: N \longrightarrow N'$

$$\begin{array}{ccc} M \otimes N & \xrightarrow{\sigma_{M,N}} & N \otimes M \\ {\scriptstyle f \otimes g} \downarrow & & \downarrow {\scriptstyle g \otimes f} \\ M' \otimes N' & \xrightarrow[\sigma_{M',N'}]{} & N' \otimes M' \end{array}$$

进一步, 下面交换图成立:

$$\begin{array}{ccc} (M \otimes N) \otimes P & \xrightarrow{a_{M,N,P}} & M \otimes (N \otimes P) \\ {\scriptstyle \sigma_{M,N} \otimes P} \downarrow & & \downarrow {\scriptstyle \sigma_{M,N \otimes P}} \\ (N \otimes M) \otimes P & & (N \otimes P) \otimes M \\ {\scriptstyle a_{N,M,P}} \downarrow & & \downarrow {\scriptstyle a_{N,P,M}} \\ N \otimes (M \otimes P) & \xrightarrow[N \otimes \sigma_{M,P}]{} & N \otimes (P \otimes M) \end{array}$$

$$\begin{array}{ccc} M \otimes (N \otimes P) & \xrightarrow{a^{-1}_{M,N,P}} & (M \otimes N) \otimes P \\ {\scriptstyle M \otimes \sigma_{N,P}} \downarrow & & \downarrow {\scriptstyle \sigma_{M \otimes N,P}} \\ M \otimes (P \otimes N) & & P \otimes (M \otimes N) \\ {\scriptstyle a^{-1}_{M,P,N}} \downarrow & & \downarrow {\scriptstyle a_{P,M,N}} \\ (M \otimes P) \otimes N & \xrightarrow[\sigma_{M,P} \otimes N]{} & (P \otimes M) \otimes N \end{array}$$

一个辫子张量范畴 $(\mathcal{C}, \otimes, I, \sigma)$ 是对称的, 如果对任意 $M, N \in \mathcal{C}$, 有: $\sigma_{N,M} \circ \sigma_{M,N} = M \otimes N$.

7.3 张 量 积

本节主要介绍在模范畴中的张量积的泛性质、存在性和相关的性质等.

设 k 是交换环, 对任意 $X, Y \in \mathcal{M}_k$. 我们有笛卡儿积 $X \times Y \in \mathcal{M}_k$, 具有结构如下:

$$(x, y) + (x', y') = (x + x', y + y'), \quad (x, y) \cdot a = (xa, ya)$$

对任意 $x, x' \in X, y, y' \in Y, a \in k$. 设 $Z \in \mathcal{M}_k$ 是第三个模, 考虑 k-双线性映射 $X \times Y \longrightarrow Z$ 的所有集合 $\mathrm{Bil}_k(X \times Y, Z)$. 那么 X 与 Y 的张量积定义为唯一的

7.3 张量积

k-模, 记为 $X \otimes Y$. 对此, 存在一个双线性映射 $\phi: X \times Y \longrightarrow Z$, 使得映射

$$\Phi_Z \operatorname{Hom}(X \otimes Y, Z) \longrightarrow \operatorname{Bil}_k(X \times Y, Z), \quad f \mapsto f \circ \phi$$

是双射, 对任意 $Z \in \mathcal{M}_k$. 张量积的唯一性由下面泛性质决定.

命题 7.3.1 如果 $T \in \mathcal{M}_k$ 带有一个 k 双线性映射 $\psi: X \times Y \longrightarrow T$, 使得对应映射

$$\Psi_Z : \operatorname{Hom}(T, Z) \longrightarrow \operatorname{Bil}_k(X \times Y, Z)$$

是双线性. 那么一定存在唯一 k-线性映射 $\tau : X \otimes Y \longrightarrow T$ 使得 $\psi = \tau \circ \phi$. $\tau = \Phi_T^{-1}(\psi)$.

我们将对张量积提供一个详细的结构. 设 $X, Y \in \mathcal{M}_k$, 且 $(X \times Y)k$ 是基为所有 $X \times Y$ 中元素标记的自由 k-模, 即 $(X \times Y)k$ 中元素是形如向量 $e_{(x,y)}$ 的有限线性组合, $x \in X, y \in Y$. 现在考虑由下面关系生成的子模 I:

$$\begin{cases} e_{(x+x',y)} - e_{(x,y)} - e_{(x',y)}, \\ e_{(x,y+y')} - e_{(x,y)} - e_{(x,y')}, \\ e_{(xa,y)} - e_{(x,y)}a, \\ e_{(x,ya)} - e_{(x,y)}a. \end{cases}$$

我们将有: $X \otimes Y = (X \times Y)k/I$. 事实上, 由 I 的定义, 存在一个 k-双线性映射

$$\phi : X \times Y \longrightarrow X \otimes Y, \quad (x,y) \mapsto e_{(x,y)}.$$

进一步, 对 $Z \in \mathcal{M}_k$, 可以定义

$$\Phi_Z^{-1} : \operatorname{Bil}_k(X \times Y, Z) \longrightarrow \operatorname{Hom}(X \otimes Y, Z), \, f \mapsto (e_{(x,y)} \mapsto f(x,y)),$$

上面映射再线性扩充. 直接可证这是 Φ_Z 的逆映射.

从现在起, 我们将记 $x \otimes y := e_{(x,y)}$, 那么 $X \otimes Y$ 中元素是有限和形式: $\sum_i x_i \otimes y_i$. 进一步, 这些元素满足下面关系式:

$$\begin{cases} (x+x') \otimes y = x \otimes y + x' \otimes y, \\ x \otimes (y+y') = x \otimes y + x \otimes y', \\ (xa) \otimes y = x \otimes (ya) = (x \otimes y)a. \end{cases}$$

如果 k 是域, 有 $\dim(X \otimes Y) = \dim(X)\dim(Y)$.

参 考 文 献

[1] Abd El-Hafez A T, Delvaux L, Van Daele A. Group-cograded multiplier Hopf (∗-) algebra. Algebra Represent. Theory, 2007, 10: 77-95.

[2] Abe E. Hopf Algebras. Cambridge: Cambridge University Press, 1977.

[3] Aguiar M. Infinitesimal Hopf algebras. Contemp. Math., 2000, 267: 1-29.

[4] Aguiar M. Infinitesimal Hopf algebras and the cd-Index of polytopes. Discrete Comput. Geom., 2002, 27: 3-28.

[5] Aguiar M, Sottile F. Structure of the Malvenuto-Reutenauer Hopf algebra of permutations. Adv. Math., 2005, 191: 225-275.

[6] Aguiar M, Sottile F. Structure of the Loday-Ronco Hopf algebra of trees. J. Algebra, 2006, 295: 473-511.

[7] Alonso Álvarez J N, Fernández Vilaboa J M, González Rodríguez R. A characterization of weak Hopf (co)quasigroups. Mediterr. J. Math., 2016, 13: 3747-3764.

[8] Alonso Álvarez J N, Fernández Vilaboa J M, González Rodríguez R, Soneira Calvo C. Projections and Yetter-Drinfel'd modules over Hopf (co)quasigroups. J. Algebra, 2015, 443: 153-199.

[9] Alonso Álvarez J N, González Rodríguez R. Crossed products for weak Hopf algebras with coalgebras splitting. J. Algebra, 2004, 281: 731-752.

[10] Andruskiewitsch N, Heckenberger I, Schneider H J. The Nichols algebra of a semisimple Yetter-Drinfel'd module. Am. J. Math., 2010, 132: 1493-1547.

[11] Andruskiewitsch N, Natale S. Tensor categories attached to double groupoids. Adv. Math., 2006, 200: 539-583.

[12] Andruskiewitsch N, Schneider H J. Lifting of quantum linear spaces and pointed Hopf algebras of order p^3. J. Algebra, 1998, 209: 658-691.

[13] Andruskiewitsch N, Schneider H J. On the classification of finite-dimensional pointed Hopf algebras. Ann. Math., 2010, 171: 375-417.

[14] Arango J A O, Tiraboschi A, Wang S H. Multiplier infinitesimal bialgebras and derivator Lie bialgebras. Preprint Southeast University and Ciudad Universitaria, 2015.

[15] Bahturin Y, Fischman D, Montgomery S. On the generalized Lie structure of associative algebras. Israel J. Math., 1996, 96: 27-48.

[16] Bahturin Y, Fischman D, Montgomery S. Bicharacter, twistings and Scheunert's theorem for Hopf algebra. J. Algebra, 2001, 236: 246-276.

[17] Beattie M, Dascalescu S, Grunenfelder L. Constructing pointed Hopf algebras by Ore extensions. J. Algebra, 2000, 225: 743-770.

[18] Beggs E, Majid S. Quasitriangular and differential structures on bicrossproduct Hopf algebras. J. Algebra, 1999, 219: 582-727.

[19] Blattner R J, Cohen M, Montgomery S. Crossed products and inner actions of Hopf algebras. Trans. Amer. Math. Soc., 1986, 298: 671-711.

[20] Böhm G. Integral theory of Hopf algebroids, Algebr. Represent. Theory, 2005, 8: 563-599.

[21] Böhm G. Hopf algebroids. Handbook of Algebra, 2009, 6: 173-235. Elsevier/North-Holland, Amsterdam.

[22] Böhm G. Comodules over weak multiplier bialgebras. Internat. J. Math., 2014, 25: 1450037, 57.

[23] Böhm G. Yetter-Drinfel'd modules over weak multiplier bialgebras. Israel J. Math., 2015, 209: 85-123.

[24] Böhm G, Gómez-Torrecillas J, López-Centella E. Weak multiplier bialgebras. Trans. Amer. Math. Soc., 2015, 367: 8681-8721.

[25] Böhm G, Nill F, Szlachanyi K. Weak Hopf algebras I: Integral theory and C^*-structure. J. Algebra, 1999, 221: 385-438.

[26] Böhm G, Szlachanyi K. Hopf algebroids with bijective antipodes: Axioms, integrals, and duals. J. Algebra, 2004, 274: 708-750.

[27] Brown R. From groups to groupoids: A brief survey. Bull. London Math. Soc., 1987, 19: 113-134.

[28] Brzeziński T. Remarks on bicovariant differential calculi and exterior Hopf algebras. Lett. Math. Phys., 1993, 27: 287-300.

[29] Brzeziński T, Militaru G. Bialgebroids, $\times_A$-bialgebras and duality. J. Algebra, 2002, 251: 279-294.

[30] Bulacu D, Nauwelaerts E. Relative Hopf modules for (dual) quasi-Hopf algebras. J. Algebra, 2000, 229: 632-659.

[31] Caenepeel S, De Lombaerde M. A categorical approach to Turaev's Hopf group-coalgebras. Comm. Algebra, 2006, 34: 2631-2657.

[32] Caenepeel S, Janssen K. Partial (co)actions of Hopf algebras and partial Hopf-Galois theory. Comm. Algebra, 2008, 36: 2923-2946.

[33] Caenepeel S, Janssen K, Wang S H. Group corings. Appl. Categ. Structures, 2008, 16: 65-96.

[34] Caenepeel S, Militaru G, Zhu S L. Frobenius and Separable Functors for Generalized Module Categories and Nonlinear Equations. New York: Springer, 2002, Lecture notes in Mathematics Vol.1787.

[35] Caenepeel S, Vercruysse J, Wang S H. Morita theory for corings and cleft entwining structures. J. Algebra, 2004, 276: 210-235.

[36] Caenepeel S, Vercruysse J, Wang S H. Rationality properties for Morita contexts associated to corings. Hopf Algebras in Noncommutative Geometry and Physics, 2005, 50(69): 113-136.

[37] Chen H X, Wang S H. Hopf modules, Miyashita-Ulbrich coactions, and monoidal center constructions. Comm. Algebra, 2002, 30: 2853-2881.

[38] Chen J Z, Wang S H. Differential calculi on quantum groupoids. Comm. Algebra, 2008, 36: 3792-3819.

[39] Chen J Z, Zhang Y, Wang S H. Twisting theory for weak Hopf algebras. Appl. Math. J. Chinese Univ. Ser. B, 2008, 23: 91-100.

[40] Chen Q G, Wang S H. Connes' pairings for a new K-theory over weak Hopf algebras. Comm. Algebra, 2013, 41: 1316-1349.

[41] Chen Q G, Wang S H. Radford's formula for generalized weak biFrobenius algebras. Rocky Mountain J. Math., 2014, 44: 419-433.

[42] Cibils C. A quiver quantum group. Comm. Math. Phys., 1993, 157: 459-477.

[43] Cibils C, Rosso M. Hopf quivers. J. Algebra, 2002, 254: 241-251.

[44] Connes A. Noncommutative Geometry. London: Academic Press, 1994.

[45] Connes A, Kreimer D. Hopf algebras, renormalization and nonommutative geometry. Comm. Math. Phys., 1998, 199: 203-242.

[46] Connes A, Kreimer D. Renormalization in quantum field theory and the Riemann-Hilbert problem I: The Hopf algebra of graphs and the main theorem. Comm. Math. Phys., 2000, 210: 249-273.

[47] Connes A, Kreimer D. Renormalization in quantum field theory and the Riemann-Hilbert problem. II: The β-funtion, diffeomorphisms and the renormalization group. Comm. Math. Phys., 2001, 216: 215-241.

[48] Connes A, Moscovici H. Hopf algebras, cyclic cohomology and the transverse index theorem. Comm. Math. Phys., 1998, 198: 199-246.

[49] Connes A, Van Daele A. The group property of the invariants of Von Neumann algebras. Math. Scand., 1973, 32: 187-192.

[50] Cuntz M, Heckenberger I. Weyl groupoids with at most three objects. J. Pure Appl. Algebra, 2009, 213: 1112-1128.

[51] Cuntz M, Lentner S. A simplicial complex of Nichols algebras. Math. Z., 2017, 285: 647-683.

[52] Dauns J. Multiplier rings and primitive ideals. Trans. Amer. Math. Soc., 1969, 145: 125-158.

[53] De Commer K. On a Morita equivalence between the duals of quantum SU(2) and quantum $E(2)$. Adv. Math., 2012, 229: 1047-1079.

[54] De Commera K, Timmermann T. Partial compact quantum groups. J. Algebra, 2015, 438: 283-324.

[55] Delvaux L. Twisted tensor product of multiplier Hopf (*-) algebras. J. Algebra, 2003, 269: 285-316.

[56] Delvaux L, Van Daele A. The Drinfel'd double versus the Heisenberg double for an

algebraic quantum group. J. Pure Appl. Algebra, 2004, 190: 59-84.

[57] Delvaux L, Van Daele A. Algebraic quantum hypergroups. Adv. Math., 2011, 226: 1134-1167.

[58] Delvaux L, Van Daele A, Wang S H. Quasitriangular (G-cograded) multiplier Hopf algebras. J. Algebra, 2005, 289: 484-514.

[59] Delvaux L, Van Daele A, Wang S H. Bicrossproducts of multiplier Hopf algebras. J. Algebra, 2011, 343: 11-36.

[60] Delvaux L, Van Daele A, Wang S H. A note on the antipode for algebraic quantum groups. Canad. Math. Bull., 2012, 55: 260-270.

[61] Delvaux L, Van Daele A, Wang S H. Bicrossproducts of algebraic quantum groups. Internat. J. Math., 2013, 24: 1250131, 48.

[62] Doi Y. Hopf extensions and Maschke type theorems. Israel J. Math., 1990, 72: 99-108.

[63] Doi Y. Braided bialgebras and quadratic bialgebras. Comm. Algebra, 1993, 21: 1731-1749.

[64] Doi, Y. Substructures of bi-Frobenius algebras. J. Algebra, 2002, 256: 568-582.

[65] Doi Y, Takeuchi M. Cleft comodule algebras for a bialgebra. Comm. Algebra, 1986, 14: 801-818.

[66] Doi Y, Takeuchi M. Multiplication alteration by two-cocycles-the quantum version. Comm. Algebra, 1994, 22: 5715-5732.

[67] Dokuchaev M, Exel R. Associativity of crossed products by partial actions, enveloping actions and partial representations. Trans. Amer. Math. Soc., 2005, 357: 1931-1952.

[68] Dong J C, Wang S H. On semisimple Hopf algebras of dimension p^2q^3. J. Algebra, 2013, 375: 97-108.

[69] Dong L H, Wang S H. Constructing new quasitriangular Turaev group coalgebras. Comm. Algebra, 2013, 41: 1217-1246.

[70] Drabant B, Van Daele A, Zhang Y. Actions of multiplier Hopf algebras. Comm. Algebra, 1999, 27: 4117-4172.

[71] Drinfel'd V G. Quasi-Hopf algebras. Leningrad Math. J., 1990, 1: 1419-1457.

[72] Drinfel'd V G. Hopf algebras and the quantum Yang-Baxter equation. Soviet Mathematics-Doklady, 1985, 32: 256-258.

[73] Etingof P, Nikshych D, Ostrik V. An analogue of Radford's formula for finite tensor categories. Int. Math. Res. Notices, 2004, 54: 2915-2933.

[74] Etingof P, Nikshych D, Ostrik V. On fusion categories. Ann. of Math., 2005, 162: 581-642.

[75] Enock M, Nest R. Irreducible inclusions of factors and multiplicative unitaries. J. Func. Analysis, 1996, 137: 466-543.

[76] Enock M, Schwartz J M. Kac Algebras and Duality of Locally Compact Groups. Berlin: Springer-Verlag, 1992.

[77] Exel R. Partial actions of groups and actions of semigroups. Proc. Amer. Math. Soc., 1998, 126: 3481-3494.

[78] Fang X L, Wang S H. Twisted smash product for Hopf quasigroups. J. Southeast Univ. (English Ed.), 2011, 27: 343-346.

[79] Fang X L, Wang S H. New Turaev braided group categories and group corings based on quasi-Hopf group coalgebras. Comm. Algebra, 2013, 41: 4195-4226.

[80] Foissy L. Finite-dimensional comodules over the Hopf algebra of rooted trees. J. Algebra, 2002, 255: 85-120.

[81] Foissy L. The infinitesimal Hopf algebra and the poset of planar forests. J. Algebr. Comb., 2009, 30: 277-309.

[82] Green J A. Left Hopf Algebras. J. Algebra, 1980, 65: 399-411.

[83] Guo Q L, Wang S H. Lax group corings. Int. Electron. J. Algebra, 2008, 4: 83-103.

[84] Guo Q L, Wang S H, Zhang Y. Inner deformations of entwined modules and their simpleness. Arab. J. Sci. Eng. Sect. C Theme Issues, 2008, 33: 205-223.

[85] Guo S J, Wang S H. Separable functors for the category of partial entwined modules. J. Algebra Appl., 2012, 11: 1250101, 15.

[86] Guo S J, Wang S H. Crossed products of Hopf group-coalgebras. Kodai Math. J., 2013, 36: 325-342.

[87] Guo S J, Wang S H. Morita contexts and partial group Galois extensions for Hopf group coalgebras. Comm. Algebra, 2015, 43 (2015): 1025-1049.

[88] Guo S J, Wang S H. Quasitriangular Turaev group coalgebras and Radford's biproduct. Algebra Colloq., 2016, 23: 427-442.

[89] Heckenberger I. The Weyl groupoid of a Nichols algebra of diagonal type. Invent. Math., 2006, 164: 175-188.

[90] Hochschild G. Cohomology and representations of associative algebras. Duke Math. J., 1947, 14: 921-948.

[91] Holtkamp R. Comparison of Hopf algebras on trees. Arch. Math. (Basel), 2003, 80: 368-383.

[92] Jimbo M. A q-difference analogue of $U(g)$ and the Yang-Baxter equation. Letters in Mathematical Physics, 1985, 10: 63-69.

[93] Johnson B E. An introduction to the theory of double centralizers. Proc. London Math.Soc., 1964, 14: 299-320.

[94] Kac G I. Ring groups and the principle of duality, I. Trans. Moscow Math. Soc., 1963: 291-339.

[95] Kac V G. Infinite Dimensional Lie Algebras. Cambridge: Cambridge University Press, 1990.

[96] Kadison L. New Examples of Frobenius Extensions. University Lecture Series, Vol. 14. Providence: A.M.S., 1999.

[97] Kan H B, Wang S H. A categorical interpretation of Yetter-Drinfel′d modules. Chinese Sci. Bull., 1999, 44: 771-778.

[98] Kauffman L H, Radford D E. Oriented quantum algebras and invariants of knots and links. J. Algebra, 2001, 246: 253-291.

[99] Kharchenko V. A quantum analogue of the Poincaré-Birkhoff-Witt theorem. Algebra and Logic, 1999, 38: 259-276.

[100] Klim J, Majid S. Hopf quasigroups and the algebraic 7-sphere. J. Algebra, 2010, 323: 3067-3110.

[101] Kochubei1 A N. Non-Archimedean duality: Algebras, groups, and multipliers. Algebra. Represent Theor., 2016, 19: 1081-1108.

[102] Koppinen M. On nakayama automorphisms of double frobenius algebras. J. Algebra, 1999, 214: 22-40.

[103] Kreimer H, Takeuchi M. Hopf algebras and Galois extensions of an algebra. Indiana Univ. Math. J., 1981, 30: 675-692.

[104] Kurose H, Van Daele A, Zhang Y H. Corepresention theory of multiplier Hopf algebras II. Int. J. Math., 2000, 11: 233-278.

[105] Kustermans J. Induced corepresentation of locally compact quantum groups. J. of Functional Analysis, 2002, 194: 410-459.

[106] Larson R, Towber J. Two dual classes of bialgebras related to the concepts of "quantum groups" and "quantum Lie algebras". Comm. Algebra, 1991, 19: 3295-3345.

[107] Li F. Weak Hopf algebras and some new solutions of the quantum Yang-Baxter equation. J. Algebra, 1998, 208: 72-100.

[108] Liu L, Shen B L, Wang S H. On weak crossed products of weak Hopf algebras. Algebra. Represent. Theory, 2013, 16: 633-657.

[109] Liu L, Wang S H. The generalized C. M. Z.-theorem and a Drinfel′d double construction for WT-coalgebras and graded quantum groupoids. Comm. Algebra, 2008, 36: 3393-3417.

[110] Liu L, Wang S H. Constructing new braided T-categories over weak Hopf algebras. Appl. Categ. Structures, 2010, 18: 431-459.

[111] Liu G H, Wang S H. A construction method for Hopf algebras. Southeast Asian Bull. Math., 2006, 30: 283-300.

[112] Liu G H, Wang S H. Graded Morita theory for group coring and graded Morita-Takeuchi theory. Taiwanese J. Math., 2012, 16 (2012): 1041-1056.

[113] Loday J L, Rono M O. Hopf algebra of the planar binary trees. Adv. Math., 1998, 139: 293-309.

[114] Loday J L, Rono M O. On the structure of ofree Hopf algebras. J. Reine Angew. Math., 2006, 592: 123-155.

[115] Longo R. A duality for Hopf algebras and for subfactors I. Commun. Math. Phys.,

1994, 159: 133-150.

[116] Lu D W, Wang S H. The Drinfel'd double versus the Heisenberg double for Hom-Hopf algebras. J. Algebra Appl., 2016, 15: 1650059, 24.

[117] Lu J H. On the Drinfel'd double and Heisenberg double of a Hopf algebra. Duke Math. J., 1994, 74: 763-776.

[118] Lu J H. Hopf Algebroids and quantum groupoids. Internat. J. Math., 1996, 7: 47-68.

[119] Ma T S, Wang S H. General double quantum groups. Comm. Algebra, 2010, 38: 645-672.

[120] Ma T S, Wang, S H. Bitwistor and quasitriangular structures of bialgebras. Comm. Algebra, 2010, 38: 3206-3242.

[121] Ma T S, Wang S H. A method to construct oriented quantum algebras. Comm. Algebra, 2010, 38: 4234-4254.

[122] Mac S L. Categories for the Working Mathematician. GTM. l5, New York: Springer, 1971.

[123] Mackenzie K. Double Lie algebroides and Second-order Geometry. I. Adv. Math., 1992, 94: 180-239.

[124] Majid S. Hopf-Von Neumann algebra bicrossproducts, Kac algebra bicrossproducts, and the classical Yang-Baxter equations. J. of Functional Analysis, 1991, 95: 291-319.

[125] Majid S. Braided momentum in the q-Poincaré group. J. Math. Phys., 1993, 34: 2045-2058.

[126] Majid S. Crossed products by braided groups and bosonization. J. Algebra, 1994, 163: 165-190.

[127] Majid S. Foundations of Quantum Group Theory. Cambridge: Cambridge University Press, 1995.

[128] Majid S. Braided-Lie bialgebras. Pacific J. Math., 2000, 192: 329-357.

[129] Masuda T, Nakagami Y. A Von Neumann algebra framework for the duality of the quantum groups. Publications of the Research Institute for Mathematical Sciences, Kyoto University, 1994, 30: 799-850.

[130] Masuoka A. The $p(n)$ theorem for semisimple Hopf algebras. Proc. Amer. Math. Soc., 1996, 124: 735-737.

[131] Menini C, Militaru G. Integrals, quantum Galois extensions, and the affineness criterion for quantum Yetter-Drinfel'd modules. J. Algebra, 2002, 247: 467-508.

[132] Milnor J W, Moore J C. On the structure of Hopf algebras. Ann. Math., 1965, 81: 211-264.

[133] Molnar R K. Semi-direct products of Hopf algebras. J. Algebra, 1977, 47: 29-51.

[134] Montgomery S. Hopf Algebras and Their Actions on Rings, CBMS Regional Conf. Series in Math., Vol. 82. Providence: A.M.S., 1993.

[135] Moscovici H, Rangipour B. Cyclic cohomology of Hopf algebras of transverse symme-

tries in codimension 1. Adv. Math., 2007, 210: 323-374.

[136] Moscovici H, Rangipour B. Hopf algebras of primitive Lie pseudogroups and Hopf cyclic cohomology. Adv. Math., 2009, 220: 706-790.

[137] Nam K B, Wang S H, Kim Y G. Explicit universal proof of $P \neq NP$. Advances in algebra towards millennium problems, 1-58, SAS Int. Publ., Delhi, 2005.

[138] Nikshych D. A duality theorem for quantum groupoids. Contemp. Math., 2000, 267: 237-243.

[139] Ng S H, Taft E J. Quantum convolution of linearly recursive sequences. J. Algebra, 1997, 198: 101-119.

[140] Nikshych D. Duality for Actions of Weak Kac Algebras and Crossed Product Inclusions of II_1 Factors. J. Operator Theory, 2001, 46: 635-655.

[141] Nikshych D. On the structure of weak Hopf algebras. Adv. Math., 2002, 170: 257-286.

[142] Nikshych D, Turaev V, Vainerman L. Invariants of knots and 3-manifolds from quantum groupoids. Topology and its Application, 2003, 127: 91-123.

[143] Nikshych D, Vainerman L. A Galois correspondence for II factors and quantum groupoids. J. Funct. Anal., 2000, 178: 113-142.

[144] Nikshych D, Vainerman L. A characterization of depth 2 subfactors of II1 factors. J. Func. Analysis, 2000, 171: 278-307.

[145] Nikshych D, Vainerman L. Finite quantum groupoids and their applications. New Directions in Hopf Algebras, in: MSRI Publications, 2002, 43: 211-262.

[146] Panaite F, Staic M D. Generalized (anti) Yetter-Drinfel'd modules as components of a braided T-category. Israel Journal of Mathematics, 2007, 158: 349-365.

[147] Panaite F, Van Oystaeyen F. L-R-smash product for (quasi-)Hopf algebras. J. Algebra, 2007, 309: 168-191.

[148] Panov A N. Ore extensions of Hopf algebras. Math. Notes, 2003, 74: 401-410.

[149] Perez-Izquierdo J M. Algebras, hyperalgebras, nonassociative bialgebras and loops. Adv. Math., 2007, 208: 834-876.

[150] Pflaum J M, Schauenburg P. Differential calculi on noncommutative bundles. Z. Phys. C, 1997, 76: 733-744.

[151] Podles P, Woronowicz S L. Quantum deformation of Lorentz group. Comm. Math. Phys., 1990, 130: 381-431.

[152] Popa S. Cocycle and orbit equivalence superrigidity for malleable actions of w-rigid group. Invent. Math., 2007, 170: 243-295.

[153] Popa S, Vaes S. Actions of $F1$ whose II factors and orbit equivalence relations have prescribed fundamental group. J. Amer. Math. Soc., 2010, 23: 383-403.

[154] Majid S. Quantum and braided Lie algebra. J. Geom. phys., 1994, 13: 307-356.

[155] Radford D E. The order of the antipode of a finite dimensional Hopf algebra is finite. Amer. J. Math., 1976, 98: 333-358.

[156] Radford D E. The structure of Hopf algebra with a projection. J. Algebra, 1985, 92: 322-347.

[157] Radford D. E. Minimal quasitriangular Hopf algebras. J. Algebra, 1993, 157: 285-315.

[158] Radford D E. On Kauffman's knot invariants arising from finite-dimensional Hopf algebras, in "Advances in Hopf Algebras". Lecture Notes in Pure and Applied Mathematics, 1994: 205-266, Dekker, New York.

[159] Radford D E. On a parameterized family of twist quantum coalgebras and the bracket polynomial. J. Algebra, 2000, 225: 93-123.

[160] Radford D E. On oriented quantum algebras derived from representations of the quantum double of a finite-dimensional Hopf algebra. J. Algebra, 2003, 270: 670-695.

[161] Radford D E. On the tensor product of an oriented quantum algebra with itself. J. Knot Theory Ramifications, 2007, 16: 929-957.

[162] Radford D E, Westreich S. Trace-like functionals on the double of the Taft Hopf algebra. J. Algebra, 2006, 301: 1-34.

[163] Reshetikhin N Y, Turaev V G. Ribbon graphs and their invariants derived from quantum groups. Comm. Math. Phys., 1990, 127: 1-26.

[164] Ringel C M. Tame Algebras and Integral Quadratic Forms. Lecture Notes in Mathematics, Vol. 1099. Berlin: Springer, 1984.

[165] Rodriguez-Romo S, Taft E J. A left quantum group. J. Algebra, 2005, 286: 154-160.

[166] Schauenburg P. Hopf modules and Yetter-Drinfel′d modules. J. Algebra., 1994, 169: 874-890.

[167] Schauenburg P. Differential-graded Hopf algebra and quantum group differential calculi. J. Algebra, 1996, 180: 239-286.

[168] Schauenburg P. Duals and doubles of quantum groupoids. New Trends in Hopf Algebra Theory, Proc. Colloq. on Quantum Groups and Hopf Algebras, La Falda, Sierra de Cordoba, Argentina, 1999, in: Contemp. Math., vol. 267, Amer. Math. Soc., Providence, RI, 2000: 273.

[169] Shen B L, Wang S H. Blattner-Cohen-Montgomery's duality theorem for (weak) group smash products. Comm. Algebra, 2008, 36: 2387-2409.

[170] Shen B L, Wang S H. On group crossed products. Int. Electron. J. Algebra, 2008, 4: 177-188.

[171] Shen B L, Wang S H. Weak Hopf algebra duality in weak Yetter-Drinfel′d categories and applications. Int. Electron. J. Algebra, 2009, 6: 74-94.

[172] Smith M F. The Pontryagin duality theorem in linear spaces. Ann. Math., 1952, 56: 248-253.

[173] Stinespring W F. Integration theorems for gauges and duality for unimodular locally compact groups. Trans. Amer. Math. Soc., 1959, 90: 15-56.

[174] Sweedler M E. Hopf Algebras. New York: Benjamin, Inc., 1969.

[175] Sweedler M E. Integrals for Hopf algebras. Ann. Math., 1969, 89: 323-335.
[176] Szlachanyi K. Finite quantum groupoids and inclusions of finite type. Fields Inst. Commun., 2001, 30: 393-413.
[177] Szymanski W. Finite index subfactors and Hopf algebra crossed products. Proc. Amer. Math. Soc., 1994, 120: 519-528.
[178] Takai H. On a duality for crossed products of C^*-algebras. J. Functional Analysis, 1975, 19: 25-39.
[179] Takesaki M. A characterization of group algebras as a converse of Tannaka-Stinespring-Tatsuuma duality theorem. American Journal of Mathematics, 1969, 91: 529-564.
[180] Takesaki M. Duality and Von Neumann algebras. Lecture Notes in Mathematics, 1972, 247: 665-785.
[181] Tatsuuma N. A duality theorem for locally compact groups. I. Proceedings of the Japan Academy, 1965, 41: 878-882.
[182] Tatsuuma N. A duality theorem for locally compact groups. II. Proceedings of the Japan Academy, 1966, 42: 46-47.
[183] Timmermann T. On duality of algebraic quantum groupoids. Adv. Math., 2017, 309: 692-746.
[184] Timmermann T, Van Daele A. Multiplier Hopf algebroids arising from weak multiplier Hopf algebras. Banach Center Publications, 2015, 106: 73-110.
[185] Turaev V. Quantum invariants of knots and 3-manifolds, de Gruyter Studies in Mathematics, 18. Walter de Gruyter & Co., Berlin, 1994.
[186] Vaes S, Van Daele A. Hopf C^*-algebras. Proc. London Math. Soc., 2001, 82: 337-384.
[187] Van Daele A. K-theory for graded Banach algebras. II. Pacific J. Math., 1988, 134: 377-392.
[188] Van Daele A. Dual pairs of Hopf *-algebras. Bulletin of the London Mathematical Society, 1993, 25: 917-932.
[189] Van Daele A. The Yang-Baxter and pentagon equation. Compositio Math., 1994, 91: 201-221.
[190] Van Daele A. Multiplier Hopf algebras. Trans. Amer. Math. Soc., 1994, 342: 917-932.
[191] Van Daele A. An algebraic framework for group duality. Adv. Math., 1998, 140: 323-366.
[192] Van Daele A. Quantum groups with invariant integrals. Proceedings of the National Academy of Sciences, 2000, 97: 541-546.
[193] Van Daele A. Tools for working with multiplier Hopf algebras. The Arabian Journal for Science and Engineering, 2008, 33: 505-527.
[194] Van Daele A, Wang S H. The Larson-Sweedler theorem for multiplier Hopf algebras. J. Algebra, 2006, 296: 75-95.

[195] Van Daele A, Wang S H. Larson-Sweedler theorem and some properties of discrete type in (G-cograded) multiplier Hopf algebras. Comm. Algebra, 2006, 34: 2235-2249.

[196] Van Daele A, Wang S H. On the twisting and Drinfel'd double for multiplier Hopf algebras. Comm. Algebra, 2006, 34: 2811-2842.

[197] Van Daele A, Wang S H. A class of multiplier Hopf algebras. Algebr. Represent. Theory, 2007, 10: 441-461.

[198] Van Daele A, Wang S H. New braided crossed categories and Drinfel'd quantum double for weak Hopf group coalgebras. Comm. Algebra, 2008, 36: 2341-2386.

[199] Van Daele A, Wang S H. Pontryagin duality for bornological quantum hypergroups. Manuscripta Math., 2010, 131: 247-263.

[200] Van Daele A, Wang S H. Weak multiplier Hopf algebras. Preliminaries, motivation and basic examples in: Proceedings of the conference "Operator algebras and quantum groups (Warsaw 2011)", Banach Center Publ. 98, Polish Academy of Sciences, Institute of Mathematics, Warszawa, 2012: 367-415.

[201] Van Daele A, Wang S H. Weak multiplier Hopf algebras I. The main theory. J. Reine Angew. Math., 2015, 705: 155-209.

[202] Van Daele A, Wang S H. Weak multiplier Hopf algebras II. The source and target algebras. ArXiv:1403.7906v2 [math.RA].

[203] Van Daele A, Wang S H. Weak multiplier Hopf algebras III. Integrals and duality. ArXiv:1701.04951v3 [math.RA].

[204] Van Daele A, Woronowicz S L. Duality for the quantum $E(2)$ group. Pacific J. Math., 1996, 173: 375-385.

[205] Van Daele A, Zhang Y. Corepresention theory of multiplier Hopf algebras I. Int. J. Math., 1999, 10: 503-530.

[206] Van Heeswijck L. Duality in the theory of crossed products. Mathematica Scandinavica, 1979, 44: 313-329.

[207] Van Oystaeyen F, Zhang P. Quiver Hopf algebras. J. Algebra, 2004, 280: 577-589.

[208] Virelizier A. Hopf group-coalgebras. J. Pure Appl. Algebra, 2000, 171: 75-122.

[209] Virelizier A. Graded quantum groups and quasitriangular Hopf group-coalgebras. Comm. Algebra, 2005, 33: 3029-3050.

[210] Voigt C. Bornological quantum groups. Pacific J. Math., 2008, 235: 93-135.

[211] Wambst M. Quantum Koszul complexes for Majid's braided Lie Algebras. J. Math. Phys., 1994, 35: 6213-6223.

[212] Wang S H. A duality theorem of crossed coproduct for Hopf algebras. Sci. China Ser. A, 1995, 38: 1-7.

[213] Wang S H. Lie bialgebra structure of multivariate linearly recursive sequences. Chinese Sci. Bull., 1996, 41: 271-275.

[214] Wang S H. On braided Hopf algebra structures over the twisted smash products.

Comm. Algebra, 1999, 27: 5561-5573.

[215] Wang S H. Quasitriangularity of the twisted smash coproduct Hopf algebras. Progr. Natur. Sci. (English Ed.), 1999, 12: 894-902.

[216] Wang S H. Braided monoidal categories of twisted smash products and the twisted smash products over braided monoidal categories. (Chinese) Acta Math. Sinica (Chin. Ser.), 1999, 42: 385-394.

[217] Wang S H. Central invariants of ρ-Lie algebras in Yetter-Drinfel'd categories. Sci. China Ser. A, 2000, 43: 803-809.

[218] Wang S H. On the braided structures of bicrossproduct Hopf algebras. Tsukuba J. Math., 2001, 25: 103-120.

[219] Wang S H. Doi-Koppinen Hopf bimodules are modules. Comm. Algebra, 2001, 29: 4671-4682.

[220] Wang S H. On the generalized H-Lie structure of associative algebras in Yetter-Drinfel'd categories. Comm. Algebra, 2002, 30: 307-325.

[221] Wang S H. A construction of braided Hopf algebras. Tsukuba J. Math., 2002, 26: 269-289.

[222] Wang S H. Braided monoidal categories associated to Yetter-Drinfel'd categories. Comm. Algebra, 2002, 30: 5111-5124.

[223] Wang S H. A generalized double crossproduct and Drinfel'd double. Southeast Asian Bull. Math., 2002, 26: 159-180.

[224] Wang S H. A Maschke type theorem for Hopf π-comodules. Tsukuba J. Math., 2004, 28: 377-388.

[225] Wang S H. Cibils-Rosso's theorem for quantum groupoids. Comm. Algebra, 2004, 32: 3703-3723.

[226] Wang S H. Group entwining structures and group coalgebra Galois extensions. Comm. Algebra, 2004, 32: 3437-3457.

[227] Wang S H. Group twisted smash products and Doi-Hopf modules for T-coalgebras. Comm. Algebra, 2004, 32: 3417-3436.

[228] Wang S H. An analogue of Kegel's theorem for quasi-associative algebras. Comm. Algebra, 2005, 33: 2607-2623.

[229] Wang S H. Morita contexts, π-Galois extensions for Hopf π-coalgebras. Comm. Algebra, 2006, 34: 521-546.

[230] Wang S H. Coquasitriangular Hopf group algebras and Drinfel'd co-doubles. Comm. Algebra, 2007, 35: 77-101.

[231] Wang S H. Turaev group coalgebras and twisted Drinfel'd double. Indiana Univ. Math. J., 2009, 58: 1395-1417.

[232] Wang S H. New Turaev braided group categories over entwinning structures. Comm. Algebra, 2010, 38: 1019-1049.

[233] Wang S H. Hopf-type algebras. Comm. Algebra, 2010, 38: 4255-4276.

[234] Wang S H. Algebraic quantum hypergroups of discrete type. Math. Scand., 2011, 108: 198-222.

[235] Wang S H. Nakayama automorphisms of generalized co-Frobenius algebras. Comm. Algebra, 2011, 39: 2449-2462.

[236] Wang S H. Kaplansky's ten conjectures. (Chinese) Adv. Math. (China), 2012, 41: 257-265.

[237] Wang S H. More examples of algebraic quantum hypergroups. Algebr. Represent. Theory, 2013, 16: 205-228.

[238] Wang S H. New Turaev braided group categories and group Schur-Weyl duality. Appl. Categ. Structures, 2013, 21: 141-166.

[239] Wang S H. Quantum convolution of binary-linearly recursive sequences. Preprint Southeast University, 2015.

[240] Wang S H. More examples of weak multiplier Hopf algebras. Preprint Southeast University, 2015.

[241] Wang S H. Twisted algebras of multiplier Hopf (*-)algebras. Preprint Southeast University, 2015.

[242] Wang S H. Pontryagin duality for algebraic quantum hypergroupoids. Preprint Southeast University, 2015.

[243] Wang S H. Quantum Lie algebras and an analogue of Berele-Regev's theorem. Preprint Southeast University, 2015.

[244] Wang S H. Mltiplier Hopf (co)quasigroups. Preprint Southeast University, 2016.

[245] Wang S H. Pontryagin duality for bornological quantum hypergroupoids. Preprint Southeast University, 2016.

[246] Wang S H. Infinite double groupoids and algebraic quantum groupoids. Preprint Southeast University, 2016.

[247] Wang S H. Multiplier cogroupoids. Preprint Southeast University, 2016.

[248] Wang S H. Partial actions of multiplier Hopf algebras. Preprint Southeast University, 2016.

[249] Wang S H. Yetter-Drinfel'd categories over Hopf (co)quasigroups. Preprint Southeast University, 2016.

[250] Wang S H, Chen H X. Hopf-Galois coextensions and braided Lie coalgebras. Progr. Natur. Sci. (English Ed.), 2002, 12: 264-270.

[251] Wang S H, Jiao Z M, Zhao W Z. Hopf algebra structures on crossed products. Comm. Algebra, 1998, 26: 1293-1303.

[252] Wang S H, Kan H B, Chen H X. Generalized H-Lie structure of associative algebras in a category. Algebra Colloq., 2002, 9: 143-154.

[253] Wang S H, Kim Y G. Quasitriangular structures for a class of Hopf algebras of di-

mension p^6. Comm. Algebra, 2004, 32: 1401-1423.

[254] Wang S H, Li J Q. H-comodule algebras, total integrals and separable extensions. (Chinese) Chinese Ann. Math. Ser. A, 1996, 17: 589-594.

[255] Wang S H, Li J Q. On twisted smash products for bimodule algebras and the Drinfel'd double. Comm. Algebra, 1998, 26: 2435-2444.

[256] Wang S H, Ma T S. Singular solutions to the quantum Yang-Baxter equations. Comm. Algebra, 2009, 37: 296-316.

[257] Wang S H, Van Daele A, Zhang Y H. Constructing quasitriangular multiplier Hopf algebras by twisted tensor coproducts. Comm. Algebra, 2009, 37: 3171-3199.

[258] Wang S H, Wang D G. Hopf algebra structure $H^{\sigma-R}$ with two sided invertible 2-cocycle. Comment. Math. Univ. Carolin., 1999, 40: 635-650.

[259] Wang S H, Zhu H X. On corepresentations of multiplier Hopf algebras. Acta Math. Sin. (Engl. Ser.), 2010, 26: 1087-1114.

[260] Wang S X, Wang S H. Drinfel'd double for braided infinitesimal Hopf algebras. Comm. Algebra, 2014, 42: 2195-2212.

[261] Wang S X, Wang S H. Hom-Lie algebras in Yetter-Drinfel'd categories. Comm. Algebra, 2014, 42: 4526-4547.

[262] Wisbauer R. Weak corings. J. Algebra, 2001, 245: 123-160.

[263] Woronowicz S L. Pseudospaces, pseudogroups and Pontrjagin duality. Proceedings of the International Conference on Mathematical Physics, Lausanne, 1979. Lecture Notes in Physics, 1980, 116: 407-412.

[264] Woronowicz S L. Compact matrix pseudogroups. Comm. Math. Phys., 1987, 111: 613-665.

[265] Woronowicz S L. Differential calculus on compact matrix pseudogroups (quantum groups). Comm. Math. Phys., 1989, 122: 125-170.

[266] Woronowicz S L. Quantum $E(2)$ group and its Pontryagin dual. Letters in Mathematical Physics, 1991, 23: 251-263.

[267] Yamanouchi T. Duality for generalized Kac algebras and a characterization of finite groupoid algebras. J. Algebra, 1994, 163: 9-50.

[268] Yang T, Wang S H. A lot of quasitriangular group-cograded multiplier Hopf algebras. Algebr. Represent. Theory, 2011, 14: 959-976.

[269] Yang T, Wang S H. Constructing new braided T-categories over regular multiplier Hopf algebras. Comm. Algebra, 2011, 39: 3073-3089.

[270] You M M, Zhou N, Wang S H. Hom-Hopf group coalgebras and braided T-categories obtained from Hom-Hopf algebras. J. Math. Phys., 2015, 56: 112302, 16.

[271] You M M, Wang S H. Constructing new braided T-categories over monoidal Hom-Hopf algebras. J. Math. Phys., 2014, 55: 111701, 16.

[272] Zhang T, Wang S H. Hopf quasimodules in Yetter-Drinfel'd quasimodule categories.

Preprint Southeast University, 2016.

[273] Zhang T, Wang S H, Gu Y. Some new braided monoidal categories over Hopf quasigroups. Preprint Southeast University, 2016.

[274] Zhang X H, Wang S H. New Turaev braided group categories and weak (co)quasi-Turaev group coalgebras. J. Math. Phys., 2014, 55: 111702, 27.

[275] Zhang X H, Zhao X F, Wang S H. Sovereign and ribbon weak Hopf algebras. Kodai Math. J., 2015, 38: 451-469.

[276] Zhang X H, Wang S H. Weak Hom-Hopf algebras and their (co)representations. J. Geom. Phys., 2015, 94: 50-71.

[277] Zhang Y. The quantum double of a coFrobenius Hopf algebra. Comm. Algebra, 1999, 27: 1413-1427.

[278] Zhao W Z, Wang S H, Jiao Z M. The Hopf algebra structure on a double crossproduct. Comm. Algebra, 1998, 26: 467-476.

[279] Zhao W Z, Wang S H, Jiao Z M. On the quasitriangular structures of bicrossproduct Hopf algebras. Comm. Algebra, 2000, 28: 4839-4853.

[280] Zhao X F, Liu G H, Wang S H. Symmetric pairs in Yetter-Drinfel'd categories over weak Hopf algebras. Comm. Algebra, 2015, 43: 4502-4514.

[281] Zhou N, Wang S H. A duality theorem for weak multiplier Hopf algebras. Internat. J. Math., 2017, 28: 1750032,34.

[282] Zhou X, Wang S H. The duality theorem for weak Hopf algebra (co) actions. Comm. Algebra, 2010, 38: 4613-4632.

[283] Zhu H X, Wang S H. A generalized Drinfel'd quantum double construction based on weak Hopf algebras. Comm. Algebra, 2010, 38: 199-229.

[284] Zhu H X, Wang S H, Chen J Z. Bicovariant differential calculi on a weak Hopf algebra. Taiwanese J. Math., 2014, 18: 1679-1712.

[285] Zhu Y. Hopf algebras of prime dimension. Int. Math. Res. Notes, 1994, 1: 53-59.

[286] Zunino M. Double construction for crossed Hopf coalgebras. J. Algebra, 2004, 278: 43-75.

[287] Zunino M. Yetter-Drinfel'd modules for crossed structures. J. Pure and Applied Algebra, 2004, 193: 313-343.

[288] 王栓宏, 陈建龙. Galois 余环理论. 北京: 科学出版社, 2009.

[289] 王栓宏. Pontryagin 对偶与代数量子超群. 北京: 科学出版社, 2011.

[290] 王栓宏. 群交叉 Yetter-Drinfel'd 范畴. 北京: 科学出版社, 2012.

[291] 王栓宏. 近世代数. 北京: 科学出版社, 2013.

名 词 索 引

B

靶, 1, 12
靶代数, 105
靶空间, 58
辫子, 1, 21
辫子张量范畴, 157, 159

C

缠绕结构, 30, 31
乘子 Hopf 代数, 1
乘子 Hopf 代数胚, 101, 122
乘子双代数胚, 123, 125
冲积, 48, 72

F

范畴, 147, 153
分次 Hopf 代数, 35, 37

H

环扩, 1, 15

J

积分, 25,
箭图, 1, 12

L

量子杨 -Baxter 方程, 64

N

拟群, 10, 87
拟三角弱 Hopf 代数, 64, 65
拟双代数, 78, 79
纽结, 1, 19

Q

群代数, 27, 32
群胚, 1, 7
群余代数, 81

R

弱 Hom-Hopf 代数, 86, 87
弱 Hopf 拟群, 89
弱 Hopf 代数, 53, 55
弱 Hopf 群代数, 82, 83
弱乘子 Hopf 代, 101, 103
弱乘子双代数, 108, 111
弱双 Frobenius 代数, 77, 78
弱双代数, 55, 57

S

深度 2 扩张, 19
双 Frobenius 代数, 76, 78
双边交叉积代数, 60
双代数, 25, 28

W

完全, 101, 124
微分 Hopf 代数, 38, 45
微积分, 135, 143
无穷小 Hopf 代, 98
无穷小乘子 Hopf 代数, 100

Y

右 Hopf 代数, 75
右双代数胚, 90, 91
余代数, 19, 28
余环, 29, 31
余拟双代数, 79, 80
源, 1, 2
源代数, 105
源空间, 58

Z

张量范畴, 157, 159
张量函子, 155, 159

左 Hopf 代数, 75, 76
左双代数胚, 89, 91

其他

Drinfel′d 量子偶, 53
Hom-Hopf 代数, 84, 85
Hopf 拟群, 87, 88
Hopf 代数, 1, 25
Hopf 代数胚, 89, 91
Hopf 方程, 141, 146
Hopf 群代数, 82, 83
H 伪代数, 96, 97
H 伪余代数, 96, 98
Long 方程, 142, 143
Pontryagin 对偶, 111, 122
Radford 双积, 48
Rota-Baxter 算子, 144, 145
Temperley-Lieb 代数, 21, 60
Yetter-Drinfel′d 模, 38, 45